GENOME REARRANGEMENT and STABILITY

SERIES EDITORS
Kay E. Davies, *MRC Clinical Sciences Centre Royal Postgraduate Medical School, London*
Shirley M. Tilghman, *Department of Molecular Biology Princeton University*

The identification and mapping of genes, analysis of their structures, and discovery of the functions they encode are now cornerstones of experimental biology, health research, and biotechnology. *Genome Analysis* is a series of short, single-theme books that review the data, methods, and ideas emerging from the study of genetic information in humans and other species. Each volume contains invited papers that are timely, informative, and concise. These books are an information source for junior and senior investigators in all branches of biomedicine interested in this new and fruitful field of research.

SERIES VOLUMES
1. Genetic and Physical Mapping
2. Gene Expression and Its Control
3. Genes and Phenotypes
4. Strategies for Physical Mapping
5. Regional Physical Mapping
6. Genome Maps and Neurological Disorders
7. Genome Rearrangement and Stability

GENOME REARRANGEMENT and STABILITY

Edited by
Kay E. Davies
MRC Clinical Sciences Centre
Royal Postgraduate Medical School, London

Stephen T. Warren
Emory University School of Medicine
Howard Hughes Medical Institute

Volume 7 / GENOME ANALYSIS

 Cold Spring Harbor Laboratory Press 1993

Genome Analysis Volume 7
Genome Rearrangement and Stability

All rights reserved
Copyright 1993 by Cold Spring Harbor Laboratory Press
Printed in the United States of America
ISBN 0-87969-388-6
ISSN 1050-8430
LC 93-72455

Cover and book design by Leon Bolognesi & Associates, Inc.

Authorization to photocopy items for internal or personal use, or the internal or personal use of specific clients, is granted by Cold Spring Harbor Laboratory Press for libraries and other users registered with the Copyright Clearance Center (CCC) Transactional Reporting Service, provided that the base fee of $5.00 per article is paid directly to CCC, 27 Congress St., Salem, MA 01970. [0-87969-387-8/93 $5.00 + .00]. This consent does not extend to other kinds of copying, such as copying for general distribution, for advertising or promotional purposes, for creating new collective works, or for resale.

All Cold Spring Harbor Laboratory Press publications may be ordered directly from Cold Spring Harbor Laboratory Press, 10 Skyline Drive, Plainview, New York 11803-2500. Phone: 1-800-843-4388 in Continental U.S. and Canada. All other locations (516) 349-1930. FAX: (516) 349-1946.

Contents

Preface *vii*

**Six Human Genetic Disorders Involving Mutant
Trinucleotide Repeats: Similarities and Differences** 1
David L. Nelson

Capturing a CAGey Killer 25
Marcy E. MacDonald, Christine M. Ambrose, Mabel P. Duyao,
and James F. Gusella

**Mechanisms of Mutations at Human
Minisatellite Loci** 43
John A.L. Armour, Darren G. Monckton, David L. Neil, Keiji Tamaki,
Annette MacLeod, Maxine Allen, Moira Crosier, and Alec J. Jeffreys

**Are Repetitive DNA Sequences Involved with
Leukemia Chromosome Breakpoints?** 59
Raymond L. Stallings, Norman A. Doggett,
Katsuzumi Okumura, A. Gregory Matera, and David C. Ward

**Genetic Control of Simple Sequence Stability
in Yeast** 79
Arthur J. Lustig and Thomas D. Petes

**Defined Ordered Sequence DNA, DNA
Structure, and DNA-directed Mutation** 107
Robert D. Wells and Richard R. Sinden

Homologous DNA Interactions in the Evolution of Gene and Chromosome Structure 139
Miroslav Radman, Robert Wagner, and Maja C. Kricker

The Use of DNA Sequence Homology and Pseudogenes for the Construction of Active VSG Genes in *Trypanosoma equiperdum* 153
Harvey Eisen and Andrew Strand

Index *161*

Preface

The origins and functions of repetitive genomic sequence have perplexed geneticists and molecular biologists for some time. However, rather than being "junk" DNA as once believed, repetitive DNA elements have emerged as key genomic sequences involved in an ever-growing number of cellular processes. This volume arose from a Banbury Conference at the Cold Spring Harbor Laboratory on DNA repeats and human gene mutations convened in the fall of 1992. This conference, and subsequently this volume, were in response to the startling finding of trinucleotide repeat expansions as a cause of human disease. Currently, six genetic disorders, including the fragile X syndrome, myotonic dystrophy, and Huntington's disease, have been found to be due to the unstable expansion of triplet repeats within exons of genes. The mechanism responsible for this unprecedented mutation remains unknown, and the chapters of this volume are designed to illuminate current knowledge in this area and, perhaps more importantly, to critically review fields of investigation that may well impinge upon repeat expansions and perhaps provide sufficient insight to develop critical experiments aimed at these mechanisms.

D.L. Nelson begins this volume with a timely review of the six disorders currently known to be due to trinucleotide repeat expansions. The similarities and differences among these disorders are well described and suggest, perhaps, a common mechanism for many. However, it remains unclear if the massive expansions seen in fragile X syndrome, myotonic dystrophy, and FRAXE mental retardation are mechanistically distinct from the more moderate expansions of Huntington's disease, spinocerebellar ataxia type 1, and Kennedy's disease. We are fortunate to have M.E. MacDonald and colleagues contribute a delightful review of Huntington's disease, created in the manner of a detective story, which contains a critical examination of the current state of research into this fascinating disorder and the responsible gene and repeat mutation.

J.A. Armour and coauthors provide an important section on minisatellite loci. These highly polymorphic sequences share some similarities with trinucleotide repeats, such as their relatively frequent rate of mutation and, in some instances, meiotic instability. Repeat instability is also the focus of the chapter by R.L. Stallings and his associates, who ask if repetitive DNA plays a role in genome rearrangement, specifically examining their role in the genesis of recurrent leukemia breakpoints. In what surely will be an important model system for the study of repeats, A.J. Lustig and T.D. Petes describe the genetic control of simple sequence stability in yeast. The ability to examine repeats against a variety of genetic backgrounds, such as mismatch repair deficiencies, coupled with clever selection assays, provides a powerful tool in the study of repetitive sequence.

R.D. Wells and R.R. Sinden examine the role of DNA structure in simple sequence instability. Ultimately, our understanding of the mechanisms of repeat instability and expansion will likely rest on the interaction of enzymatic processes with particular structure defined by the composition and length of repetitive sequence. M. Radman, R. Wagner, and M.-C. Kricker discuss the evolution of gene and chromosome structure. Clearly, repetitive elements have been maintained even though they are inherently unstable, and this chapter suggests mechanisms by which organisms cope with these sequences and what roles they may play in recombination and polymorphism. Finally, H. Eisen and A. Strand describe the curious genetic mechanism of the African trypanosome to evade host immunity, which utilizes the transposition of repeated silent genes into transcriptionally active regions. Although it is unclear if any relationship exists here with repeat expansion, it surely is similar in terms of being unusual.

We are grateful to all of the authors for their hard work in writing their contributions. We are greatly indebted to the staff of the Cold Spring Harbor Laboratory Press, especially Nancy Ford and her colleagues, Patricia Barker and Mary Cozza, who have worked so hard to ensure rapid publication.

Stephen T. Warren
Kay E. Davies
October 1993

GENOME REARRANGEMENT
and STABILITY

Six Human Genetic Disorders Involving Mutant Trinucleotide Repeats

David L. Nelson
Institute for Molecular Genetics, Human Genome Center
Baylor College of Medicine, Houston, Texas 77030

This chapter compares and contrasts several aspects of the six human genetic disorders caused by trinucleotide repeat expansion. The six disorders are fragile X syndrome (FraX), mutation resulting in a closely related fragile site (FRAXE), myotonic dystrophy (DM), spinobulbar muscular atrophy (SBMA), Huntington's disease (HD), and spino-cerebellar ataxia type 1 (SCA1). Each is discussed with regard to phenotype variability and its dependence on the nature and behavior of the underlying mutant repeat sequences. Where known, the nature of the gene products and the effects of the mutations on them are presented. Aspects of the genetics of these disorders (in particular, penetrance, expressivity, and anticipation) are related to observations of the repeat sequence mutations and their characteristics. Speculation on the effects of similarly unstable mutations on genetic analysis of complex phenotypes is offered.

Main points discussed include:

❑ phenotypes of the six disorders
❑ genetics of the disorders
❑ repeat characteristics and behavior
❑ genes and gene products
❑ trinucleotide instability and other genetic mechanisms

INTRODUCTION

In just two years, the efforts of positional cloning aided by the human genome initiative have uncovered six novel human mutations involved in genetic disorders with peculiar genetic features. The feature common to these six mutations is the presence of a mutation-prone sequence taking the form of a GC-rich trinucleotide repeat. Although they may be different on the coding strand, two basic repeat sequences (CAG and CGG) have been found to be involved. Three of the disorders are among the most common of human genetic diseases, suggesting that this mutational mechanism may be responsible for significant human genetic morbidity. The history of the discovery of the gene defects in each of the disorders is quite interesting but beyond the scope of this chapter. A brief chronology of the disease discoveries follows.

Mutations in FraX-related fragile site (FRAXA) were described in three articles in May of 1991 (Oberlé et al. 1991; Verkerk et al. 1991; Yu et al. 1991). Two of these described a CGG trinucleotide repeat potentially involved in the large expansion mutations observed by Southern hybridization, but it was a fourth article (Kremer et al. 1991) which demonstrated the direct involvement of the trinucleotide repeat in the expansion mutations. At the same time, attempts to determine the potential involvement of the androgen receptor in an X-linked muscular atrophy known as SBMA had uncovered a heterogeneous series of mutations in a previously described polymorphic region of the protein encoded by a trinucleotide repeat of the form CAG (Edwards et al. 1991; La Spada et al. 1991). Although these two disorders were found to involve mutations in trinucleotide repeats, there were sufficient differences between them (see below) to suggest that this was only coincidence. However, the unusual features of the genetic transmission of FraX suggested candidate human diseases for similar mutational mechanisms (Sutherland et al. 1991). The leading candidate was the gene defect in DM, and groups attempting to localize that disorder began to use the trinucleotide paradigm.

Early in 1992, the genetic lesion in DM was identified as a CAG repeat by several interrelated groups (Brook et al. 1992; Fu et al. 1992; Harley et al. 1992; Mahadevan et al. 1992). The repeat showed behavior quite similar to that in FraX, with some significant differences. Both FraX and DM were found to be very different from SBMA in most features. In October of 1992, a meeting was held at the Banbury Center of Cold Spring Harbor Laboratory to discuss the similarities and differences between these three disorders, as well as to bring together researchers with expertise in simple repeats in other systems. That meeting is responsible for this volume of articles. However, since that time, three additional disorders involving trinucleotide mutations have been discovered (Huntington's Disease Collaborative Research Group 1993;

Knight et al. 1993; Orr et al. 1993). One of these (HD) has a chapter devoted to it in this volume, and was discovered in the spring of 1993. Summer of 1993 saw the discovery of a CAG repeat mutation in SCA1 and a GCC repeat mutation in an X chromosome fragile site associated with mental retardation and located slightly distal to FRAXA on the human X chromosome, FRAXE. As described below, the mutations found in SCA1 and HD show significant similarity to the mutation found in SBMA.

Although all six disorders share the common feature of trinucleotide repeat mutation, striking differences among them suggest that this mutational mechanism constitutes their primary similarity. The consequences of each mutation must be understood separately, because the differences between them are substantial. This is not surprising when one considers other mutational mechanisms, but there has been a tendency to treat the trinucleotide disorders as being quite similar. This is likely due to the novelty of the disorders and will diminish with time; such comparisons are not made between all disorders caused by deletions, for example.

In this chapter, I seek to enumerate the similarities and differences among these six disorders in order to provide a snapshot of these interesting and evolving genetic elements. It is much too soon to know many of the details of each disorder, but there are sufficient data regarding the medical, genetic, and biochemical aspects of the diseases for a cogent discussion of each. Additional review will be appropriate as new disorders are discovered and as the many questions are answered in this new category of human mutation.

PHENOTYPES

Each of these six disorders involves cells of the neuromuscular system. In fact, with the single exception of DM, where muscle appears to be the primary tissue affected (in addition to neurons), all the disorders primarily involve neurons. In the two fragile-site-associated disorders, mental retardation is the hallmark of the phenotype; in the case of the *FMR1* gene at FRAXA, neurons show high levels of expression (Abitbol et al. 1993; Devys et al. 1993; Hinds et al. 1993; Verheij et al. 1993). The phenotypes of HD, SCA1, and SBMA show the most striking similarities. Each is due to loss of particular subsets of neurons in an age-dependent process. In HD, neurons in the striatum are lost, resulting in chorea and eventual decline of cognitive functions. In SCA1, neurons of the inferior olive in the cerebellum and of the brain stem are lost, resulting in ataxia and motor weakness. In SBMA, neurons of the dorsal root ganglia and motor neurons in the spinal and bulbar roots are affected, resulting in muscular atrophy.

Age of onset

Each of these latter three disorders usually involves variable onset sometime in adulthood. Rare juvenile-onset cases are found, and the spread in age of onset can be considerable. In HD and SCA1, for example, the average age of onset of the disease is in the late 30s, but onset can vary from under 5 years to nearly 70. Juvenile cases are much more often found among offspring of affected fathers than affected mothers in both disorders (Farrer et al. 1992; Orr et al. 1993). Correlation of age of onset with the number of CAG repeats in these disorders suggests a direct role of the number of glutamine residues in each protein in triggering onset of disease (Igarashi et al. 1992; La Spada et al. 1992; Andrew et al. 1993; Duyao et al. 1993; Orr et al. 1993; Snell et al. 1993) (see below). It is interesting that each of the late-onset neurodegenerative disorders caused by increased numbers of CAG repeats is due to death of aged, postmitotic cells. DM age of onset correlates with repeat number as well (Brook et al. 1992; Tsilfidis et al. 1992; Harley et al. 1993; Redman et al. 1993); however, the number of CAG repeats in the myotonin protein kinase gene also correlates with the severity of the phenotype.

Variability in phenotypes

Variability of the phenotype among closely related family members is another common feature of these disorders. The variable expressivity in HD, SCA1, and SBMA is represented by the age of onset, as described above. In FRAXA, however, the degree of mental impairment can be highly variable, and in extreme examples, incomplete penetrance in this syndrome leads to affected and unaffected male sibs inheriting the same at-risk genotype in the same generation (Sherman et al. 1984, 1985). Females are often found divergent in phenotype, and this is likely due to effects of X-inactivation (Rousseau et al. 1991). In addition, a constellation of phenotypic defects can be found in FraX, ranging in prevalence from 10% to 90% of patients (Hagerman 1991). Thus, expressivity can be variable as well.

The example of increasing expressivity with subsequent generations inheriting DM is striking enough to be the basis of the anticipation definition (Harper et al. 1992). Here, extreme variability in phenotype (from late-onset cataracts to a lethal congenital form) is seen, with later generations exhibiting larger numbers of repeats and more severe forms. In FRAXA and DM, significant mosaicism is found in the repeat size within cells of a single individual. This likely accounts for variable expression in these diseases, as the mutations can vary from tissue to tissue, allowing the potential for the variety of pathogenic effects observed.

This discussion of phenotype does not include the clinical features associated with the FRAXE mutation, since the phenotypic aspects of this disorder are much less well developed. The distinction between FRAXA and FRAXE was only recognized recently with the development of probes at the FRAXA locus (Sutherland and Baker 1992). Families exhibiting FRAXE were found among those previously considered to have FraX. However, these families did not exhibit the CGG expansion at FRAXA, despite the presence of a fragile site in the appropriate Xq27.3 band. Molecular probes allowed the two fragile sites to be distinguished cytogenetically using fluorescence in situ hybridization, with the finding that the FRAXE site was some 500 kbp distal to FRAXA. This allowed families carrying FRAXE mutations to be sorted from those with FRAXA, but remained a difficult assay. With the recent identification of the expanded GCC repeat at the FRAXE locus, it is now possible to unequivocally identify families carrying this mutation (Knight et al. 1993). As a consequence, the development of a clinical definition of this disorder will accelerate. Currently, the disorder is associated with mild mental retardation in males, but no other pathognomic features are known. It is still not clear whether the mental retardation observed is due to the mutation or represents a coincident finding due to bias of ascertainment (Sutherland and Baker 1992).

GENETICS

Penetrance, expressivity, and anticipation

A characteristic linking several of the diseases is deviation from standard Mendelian inheritance. This is most pronounced in FraX, where nonpenetrant individuals have penetrant offspring, and the risk of this occurrence increases in subsequent generations. In the other disorders, it is the severity of the disease which varies (variable expressivity; see above) and the trend toward earlier onset or more severe disease has been termed anticipation. This phenomenon is most clearly observed in DM.

Three of the mutations are found on the human X chromosome (FRAXA, FRAXE, and SBMA), and the others are autosomal (DM, chromosome 19; HD, chromosome 4; and SCA1, chromosome 6). Most segregate in a dominant fashion, with the two exceptions being the X-linked recessives SBMA and FRAXE. In the case of FraX, dominant segregation is demonstrated by the finding of affected female carriers. However, the penetrance of this phenotype in females is reduced to approximately 30% (50% of full mutation carriers and none of the premutation carriers; see below), and this effect is likely due to cell autonomy of the protein product coupled with X-inactivation of the normal gene. Autosomal dominance is found in the other three disorders, and this fits with the proposed mechanism of action in HD and

SCA1. Dominant inheritance is somewhat more difficult to reconcile with the mutation in DM, as is discussed below.

One of the more interesting aspects of the trinucleotide repeat mutations is their capacity for change. Each of these mutations confers a mutability onto itself, and the term "dynamic mutations" has been coined to describe this phenomenon (Richards and Sutherland 1992). The number of repeats is found to change frequently upon transmission from parent to child in each disorder, with larger repeats showing a trend of greater magnitude shifts in intergenerational changes. Mosaicism is found in FRAXA, FRAXE, and DM among the largest mutations (termed "full" mutations in FRAXA) in blood and other tissues isolated from single individuals. Thus, these large expanded alleles are mitotically unstable. No evidence for somatic mosaicism has been found in HD, SCA1, or SBMA, but these generally represent much smaller expansions. The dynamic nature of the repeats offers a logical explanation for the incomplete penetrance, variable expressivity, and anticipation found in many of these disorders, as the mutations differ within and between individuals within a family. Differing mutations can readily explain differing phenotypes.

Sex-specific effects on transmission of the mutation

A paradoxical finding in FRAXA was first described by Sherman and co-workers (Sherman et al. 1984, 1985). In this disease, two peculiar features of penetrance are noted. The probability of mental retardation for any given individual in a fragile X pedigree is determined by the affected status of close relatives, and daughters of carrier males are never at risk for disease, whereas their grandsons and granddaughters are at significant risk after passage through the female. The probability of having an affected child is dependent on the number of repeats carried by the mother (Fu et al. 1991; Heitz et al. 1992; Yu et al. 1992), with repeats tending to increase in length in subsequent generations. This provides an explanation for the increasing penetrance in later generations. The requirement of passage through the female germ line to facilitate expansion into sizes resulting in phenotypic consequences is not yet understood. Observations indicate that paternal transmissions of the FRAXA mutation never result in repeats greater than 200 in length, even when transmitted from affected males (Willems et al. 1992). Recent analysis of sperm from affected males indicates that mutations are restricted to the smaller "premutation" size despite "full" mutations (>200 repeats) in somatic cells (Reyniers et al. 1993). At first blush, this would offer an explanation for the lack of affected daughters of carrier males: Selection in sperm or their precursors eliminates cells containing the larger numbers of repeats. However, the problem of generating such males exhibiting full mutations in every somatic tissue examined but with premutations in germ-line cells suggests a postconceptional increase in repeat number in somatic tissues (Reyniers et al. 1993). These males would arise by fertilization of a premutation-containing egg. A mechanism by which the maternally contributed X chromosome is capable of expanding whereas the paternally contributed chromosome cannot is required to support this hypothesis (Nelson and Warren 1993).

Similar germ-cell-specific events are important in the other disorders. Congenital DM is inherited exclusively from mothers (Harper 1989) and is associated with very large expansion mutations, yet (unlike FRAXA) large expansions are found in both male and female transmissions of the DM mutation (Redman et al. 1993). The exclusively maternal transmission of congenital DM may reflect a maximum allowable amount of increase in alleles transmitted by males, which is less than that necessary to cause the congenital form. It is unlikely to be accounted for by parental imprinting (Jansen et al. 1993). As described above, the juvenile-onset forms of HD and SCA1 are typically inherited from fathers. In the case of HD, increased instability of the mutant allele has been observed in sperm samples from affected males, and paternal transmissions involve larger increases than maternal transmissions (Duyao et al. 1993). Since juvenile-onset cases are found to involve the largest numbers of repeats, the larger increases found in sperm than in eggs would tend to confine juvenile cases to offspring of affected fathers.

Linkage disequilibrium

There is significant linkage disequilibrium found in DM, FRAXA, and HD (MacDonald et al. 1992; Richards et al. 1992; Imbert et al. 1993; Oudet et al. 1993). This common finding is at odds with expectations in several respects. FRAXA was long considered a disorder of new mutation origin, with the frequency of new mutation among the highest known (Nussbaum and Ledbetter 1989). Since affected males rarely reproduce, the FRAXA mutation is a genetic lethal. Yet the disorder is very common (~1/1500 males). In order to maintain a high population frequency, a high new mutation rate is required. Linkage disequilibrium is inconsistent with a high frequency of new mutations, since these would likely arise on many different chromosome backgrounds. The findings in DM are instructive in this regard. Disequilibrium is well established in this disorder (Harley et al. 1991), yet apparently new mutations can arise within a few generations. This leads to the conclusion that there is a predisposed chromosome. Indeed, recent studies suggest that the chromosome at risk in DM is the one containing the longer normal alleles (Imbert et al. 1993), which are predisposed to mutate to longer alleles, which are either pathological themselves or at increased risk for further expansion. This threshold-related mechanism is likely at work in FRAXA and HD as well.

Until recently, HD was considered a "no new mutation" disease; however, genetic analysis of mutant chromosomes found a common haplotype background for only one of three HD chromosomes (MacDonald et al. 1992). Here again, it is likely that an at-risk haplotype with a normal number of repeats at the high end of the scale will be found. Studies of the normal population suggest that a small fraction of chromosomes exhibit repeat sizes verging on the size found in HD chromosomes (Andrew et al. 1993; Duyao et al., 1993; Huntington's Disease Collaborative Research Group 1993; Snell et al. 1993). Although the nature of the at-risk chromosome in FRAXA is not yet established, there is a substantial pool of chromosomes with high-end normal repeat numbers, some of which exhibit intergenerational instability (Fu et al. 1991). These are likely to persist for many generations without deleterious effect, accounting for the disequilibrium observed (Chakravarti 1992; Morton and Macpherson 1992). It remains to be seen whether disequilibrium can be established for SCA1, SBMA, or FRAXE.

REPEAT CHARACTERISTICS AND BEHAVIOR

Position

The repeat sequences and their behavior vary considerably. In common are the nature of the repeats (GC-rich and trinucleotide) and their increasing instability with increasing length, with similar levels of instability found at similar sizes (Table 1). Differences are numerous, however. Three of the six (DM, FRAXA, and FRAXE) exhibit very large increases (severalfold) in length, which can occur in a single intergenerational transmission. The other three are not found to exceed approximately 100 repeats (less in SBMA) and rarely increase by more than 10 repeats in a transmission. This difference likely relates to the position of these latter repeats within the protein-coding regions of their respective genes, although it is not understood at what level such a constraint is operating, as examples of loss-of-function mutations in these genes have not demonstrated severe phenotypic consequences. It is possible that large expansions within these genes are lethal in early embryogenesis (or in the generation of sperm and egg) and have not been recovered. This could be due to extreme gain of function similar to that conferring lethality in the selected neuronal loss found later in life. Of the three repeats capable of expanding to large sizes, FRAXA repeats are found in the 5'untranslated region of the *FMR1* gene (Verkerk et al. 1991), whereas DM repeats are found in the 3'untranslated region of the DM-kinase gene (Brook et al. 1992; Fu et al. 1992). Oddly, each is transcribed, suggesting a functional role of the repeats at the mRNA level. No gene has yet been identified for FRAXE.

Table 1 Characteristics of human genetic disorders involving mutant trinucleotide repeats

Disease	Repeat seq. amino acid (parent)	Stable normal range	Unstable premutation range	Unstable disease range	Repeat location	Methyl-ation	Nature of instability	Functional change
FRAXA	CGG n/a (CCG)	6 to ~50	~45 to ~200	>230 to >2000	1st exon 5' untranslated	++	germ line and somatic	loss
FRAXE	GCC ?? (CCG)	6 to 25	116 to 133?	>200 to >850	??	++	germ line and somatic	??
DM	CTG n/a (AGC)	5 to 37	–	50 to >2000	last exon 3' untranslated	–	germ line and somatic	altered mRNA stability?
SBMA	CAG Gln (AGC)	12 to 34	–	40 to 62	1st exon translated	–	germ line	gain
HD	CAG Gln (AGC)	11 to 36	–	42 to 100	1st exon translated	–	germ line	(gain)
SCA1	CAG Gln (AGC)	19 to 36	–	43 to 81	exon translated	–	germ line	(gain)

This table lists some of the features regarding the six genetic disorders caused by mutations in GC-rich trinucleotide repeats. The column labeled Repeat seq. provides the nucleic acid repeat sequence in the reading frame of the mRNA (where applicable), followed by the amino acid sequence repeated and the parental triplet repeat (to illustrate the common origins). Size ranges from normal unaffected individuals (normal), those at risk of having affected offspring (premutations), and those affected by the disorder (disease) are given in number of triplet repeats. Methylation of neighboring sequences is seen in expanded repeats for both FRAXA and FRAXE. Question marks indicate unknown or preliminary data. Under Functional change, both HD and SCA1 are listed as gain-of-function mutations in parentheses as there is no definitive evidence of this, but it is the presumption of the author and other workers in the field.

Small-scale changes

It is intriguing that intergenerational instability becomes apparent at roughly the same number of repeats in each disorder (Table 1). This suggests a common mechanism for conferring the small-scale instability seen in each mutation. The large-scale increases, apparently limited to FRAXA, FRAXE, and DM, are likely mediated by a process distinct from the small-scale instability. Each of the trinucleotide repeats can be expressed in the form CPuG, where Pu indicates a purine (A or G) residue. Structural aspects of this sequence (see Wells and Sinden, this volume) may play a significant role in the small-scale instability observed in families, especially since these events are common to all repeats and begin at roughly the same threshold (~40 repeats or 120 bp). Difficulty in amplification of these regions using a variety of DNA polymerases suggests replication and/or repair as a likely source of mutation in the small-scale range (Caskey et al. 1992; Strand et al. 1993). The degree of small-scale intergenerational instability appears related to the size of the repeat, but it may also vary with the location of the repeat. FRAXA premutations (typically >60 repeats) show essentially 100% instability (Fu et al. 1991), whereas mutant SBMA alleles (mostly between 40 and 50 repeats) alter in approximately one-third of transmissions (Biancalana et al. 1992; La Spada et al. 1992). In HD (38–50 repeats), approximately 80% of mutant alleles are found to change size upon transmission (Duyao et al. 1993).

Large-scale expansions

The large-scale expansions are found for both types of repeats (CGG in FRAXA and FRAXE and CAG in DM), suggesting a process independent of the precise sequence of the trinucleotide. The large expansions in FRAXA and FRAXE both result in a chromosomal fragile site; however, no such site has been associated with large DM expansion. Differences in transmission and expansion are seen between FRAXA and DM, however, with large-scale FRAXA expansions occurring exclusively on the maternally contributed X chromosome and large-scale DM expansions capable of occurring on both maternal and paternal chromosomes 19. The likelihood of large-scale expansion increases with the size of the mutant allele in both DM and FRAXA (Fu et al. 1991; Barceló et al. 1993). In FRAXA, the likelihood of large-scale expansion is greater than 99% for alleles of 90 repeats or more. Alleles of this size are very scarce in HD and SCA1 and result in juvenile onset of the disorder with little likelihood of transmission. This might explain why no large expansions have been observed in these repeats. In SBMA, the largest mutant allele contains 62 repeats (La Spada et al. 1992), a length unlikely to undergo large expansion in FRAXA.

Reduction

Contractions in repeat number are observed, but these are heavily outweighed by expansions among small-scale changes. Thus, there is a tendency for the mutations to grow with transmission to the next generation, a feature necessary to explain the anticipation observed. Among the large alleles in FRAXA, FRAXE, and DM, rare contractions into the normal size range have been observed in DM (Shelbourne et al. 1992; Abeliovich et al. 1993; Brook 1993; O'Hoy et al. 1993). The situation is less clear in FRAXA. A substantial percentage of male FraX patients (~20%) exhibit mosaicism including alleles of the premutation and full mutation varieties (Rousseau et al. 1991). These patients' premutation alleles usually comprise 10–20% of the DNA extracted from blood. One interpretation of the genesis of this type of mosaicism is that the full mutation allele reduces in size to a premutation in some cell lineages, demonstrating the ability of the full mutation to reduce (oddly, these potential reductions are always found to be premutation in size, never back to the normal length). The alternate interpretation is that the expansion to full mutation occurs postzygotically in many tissues but is frequently incomplete, leaving premutation alleles in some lineages. This latter view is supported by the presence of premutation alleles in sperm of full mutation males (Reyniers et al. 1993). Under this second hypothesis, the full mutation alleles at FRAXA are very stable and unlikely to reduce, a view supported by experimental evidence from twin studies and cells in culture (Devys et al. 1992; Wöhrle et al. 1993). However, if full mutation expansions are confined to somatic cells, these alleles may never transmit through the germ line, where reductions might occur.

Normal function

A feature common to all of the trinucleotide repeats is that they are present on the mRNA produced by each gene. In the cases of the CAG repeats in HD, SCA1, and SBMA, each encodes a run of glutamine residues in its protein product. These runs of glutamine are found to be quite polymorphic in the human population; however, the extremes of normal variation in each are remarkably similar (Table 1), with a low of 11, 12, or 19, and highs of 36, 34, and 36 in HD, SBMA, and SCA1, respectively. It is difficult to reconcile the degree of polymorphism in these regions with a vital function in the protein. Other proteins with polyglutamine stretches have been characterized, many from *Drosophila* developmental mutations affecting the nervous system (Wharton et al. 1985; Vaessin et al. 1991; Bellen et al. 1992), and transcription factors (Courey et al. 1989). The nature of action of polyglutamine stretches is unknown, although they may form spacers between protein domains (Benson and Pirrotta 1988), a view consistent with the extensive polymorphism observed. The nucleic acid sequences encoding these repeats of amino acids found in other genes do not have long stretches of the same codon; there is usually variation in the codon used.

Of some interest is the finding of conservation of the repeat sequences in FRAXA and DM in other species of mammals (Jansen et al. 1992; Ashley et al. 1993b). This suggests a functional significance that has been conserved during evolution even though the repeats in these genes are outside of the protein-coding sequences. Recent evidence indicates that nuclear proteins exist in human cells which specifically bind trinucleotide repeats (Richards et al. 1993), further suggesting that the repeats are recognized in DNA and/or RNA forms and perform a function related to gene regulation or chromatin structure.

GENES AND GENE PRODUCTS

One of the largest areas of difference in the six disorders is in the nature of their gene products. This is the area where the largest divergence will likely be found in the six disorders to date. Although trinucleotide repeats form the basis of the mutational changes in these genes, there is no reason to expect all the gene products to have similar functions (no similarity is noted between the gene products outside of the repeat sequences). Nor is it likely that a common mechanism of perturbation of gene function by the expanded number of repeats will be found. Indeed, this is already clear from studies of FRAXA and DM (see below).

HD, SCA1, and SBMA

The situations in HD, SCA1, and SBMA may be different, as each of these additions to the number of glutamines acts in a dominant manner consistent with gain of function by the encoded protein. Loss-of-function mutations in the androgen receptor exhibit testicular feminization (La Spada et al. 1991), a very different phenotype. Clear loss-of-function mutations in the genes involved in HD and SCA1 have not yet been reported. Curiously, the affected cell types (neurons) in each of these three disorders are similar, but the regional variability in loss of cells is unexplained. This is particularly difficult to understand in the case of HD, a ubiquitously expressed gene with very limited pathology in its mutant form. This difference suggests that cell-specific partners must be involved to mediate the narrow toxicity of each mutant product. These partners are likely to be affected by the increased number of glutamine residues in the various proteins, although it is formally possible that the effect is mediated through the CAG repeats themselves in either DNA or RNA form. Proof awaits the creation of model systems. The androgen receptor is well studied and has a defined normal function involving recognition of androgens and gene regulation through DNA binding. The HD gene product has no known function and no significant similarities to other known proteins outside of the polyglutamine and polyproline regions (Huntington's Disease Collaborative Research Group 1993). The product of the SCA1 locus is currently being characterized.

It is interesting to note that each of the repeats in HD, SCA1, and SBMA encodes a polyglutamine (CAG) sequence, when the repeat involved (on the upper strand) could encode polyserine (AGC) or polyalanine (GCA). On the lower strand, the repeat could produce polyalanine (GCT), polyserine (TGC), or polyleucine (CTG). Yet each of the proteins involved in these disorders contains polyglutamine. This suggests quite strongly that it is the amino acid sequence involved that is mediating the effect, rather than the repeated DNA or RNA sequences. It seems likely that some commonality of function (or misfunction) of the polyglutamine stretches will be found among these three proteins. Studies of the interactions of the polyglutamine sequences in these proteins will likely elucidate the toxicity observed. A very intriguing hypothesis involving misrecognition of the mutant HD, SCA1, and SBMA protein products by brain-specific transglutaminases has recently been proposed to account for many of the features common to these disorders (Green 1993).

FMR1

The situation in FraX is the clearest to date. The CGG repeats reside in a region of the *FMR1* gene just 3' to the promoter and are transcribed but not translated into protein. An ATG start codon just 3' to the repeats initiates protein synthesis, and no size variation is observed in protein products produced from genes with different lengths of repeats (Ashley et al. 1993a; Devys et al. 1993; Siomi et al. 1993; Verheij et al. 1993; D.P.A. Kuhl et al., in prep.). Expansion of the CGGs to lengths beyond approximately 230 repeats is usually accompanied by methylation of the CpG island at the promoter and of the repeats themselves (Bell et al. 1991; Oberlé et al. 1991; Pieretti et al. 1991; Rousseau et al. 1991; Vincent et al. 1991; Hansen et al. 1993; Hornstra et al. 1993; Richards et al. 1993). Methylation correlates well with loss of transcription of the *FMR1* gene, although a causal relationship has not been established (Pieretti et al. 1991; Sutcliffe et al. 1992). Absence of mRNA results in loss of the *FMR1* protein product, which correlates with pathology. A recent study of brothers with borderline full mutations found good correlation with the extent of methylation of the CpG island and phenotypic consequences in alleles with similar numbers of repeats (McConkie-Rosell et al. 1993). The most parsimonious interpretation of these data suggests that expansion to full mutation alleles is accompanied by methylation and loss of transcription. Two deletion mutations found in males with phenotypes congruent with FraX support a loss-of-function mechanism (Gedeon et al. 1992; Wöhrle et al. 1992). This is consistent with the observation of approximately 50% affected females among carriers of full mutations (Rousseau et al. 1991), as skewed X-inactivation can account for a loss-of-function mutation. A very similar mechanism likely exists in FRAXE, where methylation is also found associated with large mutations (Knight et al. 1993), although the affected gene is not identified.

The observation of loss of transcription of the *FMR1* gene in fragile-X-affected individuals, coupled with the identification of deletions of substantial size in two patients, suggested that the *FMR1* gene may be flanked by other genes contributing to the phenotype. It is conceivable that CGG-induced methylation spreads to adjacent genes, down-regulating their expression as well, and that these adjacent genes are co-deleted in the other mutations. The singular role of *FMR1* has been demonstrated by a point mutation in a severely affected fragile X male (De Boulle et al. 1993). This individual demonstrates that mutation of *FMR1* is sufficient for elaborating the fragile X phenotype.

Recent work has begun to unravel the function of the *FMR1* gene product (FMRP). Identification of two conserved domains in the predicted FMRP amino acid sequence similar to domains found in families of RNA-binding proteins led to experiments by two groups to define the capacity of FMRP to bind RNA (Ashley et al. 1993a; Siomi et al. 1993). The observation of RNA binding coupled with the finding that the majority of FMRP is found in the cytoplasm of expressing cells suggests a function involving RNA transport or sequestration (Devys et al. 1993; Verheij et al. 1993; D.P.A. Kuhl et al., in prep.).

DM kinase

The activity of the gene mutant in myotonic dystrophy (DM kinase) as a protein kinase of the serine/threonine type is rather clear from protein similarity studies (Brook et al. 1992; Fu et al. 1992). However, the nature of the effect of expanding the number of CTG residues in the 3' untranslated region of this mRNA remains mysterious. The dominant nature of the inheritance of DM suggests another gain-of-function mutation, and several reasonable hypotheses were put forward, including overexpression of the kinase mediated by increased mRNA stability due to lengthening of the 3' untranslated region (Brook et al. 1992; Fu et al. 1992). However, conflicting data in studies of expression of the mutant form of the DM kinase gene do not allow acceptance of any of the hypotheses as yet (Fu et al. 1993; Hofmann-Radvanyi et al. 1993; Sabouri et al. 1993). Although it seems most likely that this mutation exerts its effect on mRNA and protein levels, the possibility of effects at the chromosomal DNA level (e.g., on adjacent genes, or by eliciting aberrant transcription) cannot be ruled out at this time.

At least three different effects are found as a consequence of trinucleotide repeat mutations. These are (1) CAG expansion leading to increased length of polyglutamines, (2) CGG expansion leading to methylation and loss of expression, and (3) CTG expansion leading to perturbation of mRNA stability. This list is likely to grow as additional mutations in trinucleotide repeats are uncovered.

TRINUCLEOTIDE INSTABILITY AND OTHER GENETIC MECHANISMS

The following passage from a human genetics text illustrates the thinking about penetrance and expressivity well before the discovery of unstable DNAs:

> It may be wondered whether variable expression of a gene may not be caused by variability of the gene itself—that is, mutation—rather than by variability in the network of reactions. This is not impossible, but the probability that it is true is very small. Stability or instability of genes has been ascertained by study of their transmission through germ cells from generation to generation. Such studies have established that genes at any given locus are transmitted unchanged in the overwhelming majority of all gametes, an observable mutation occurring in usually less than one of 50,000 cells. We have little justification for believing that the genes in body cells are less stable than those in germ cells. It is therefore not reasonable to explain the high variability of certain phenotypic effects by a low stability of the genes that control them (Stern 1973).

Once the dogma that all DNA is intrinsically stable is relaxed, it becomes reasonable to consider mutations conferring instability onto themselves in areas of genetics (beyond penetrance and expressivity) with significant deviation from Mendelian expectations. Clearly, DNA instability does not account for all variation from complete penetrance, nor does it always play a role in variable expression of a phenotype. However, it does offer a simple, molecular explanation for these phenomena, and it provides a possible handle on identification of disease genes exhibiting such variability. The genome may well be much more dynamic than previously appreciated. Recent studies of murine and human micro- and minisatellites have demonstrated high mutation frequencies in these sequences as well (Kelly et al. 1989, 1991; Gibbs et al. 1993; Weber and

Wong 1993). In addition, there may be genes controlling the stability of such sequences which can mutate to increase the frequency of mutation in somatic tissues (Aaltonen et al. 1993; Strand et al. 1993; Thibodeau et al. 1993).

Multifactorial inheritance

As first suggested by Sutherland and colleagues in their prescient paper (Sutherland et al. 1991), many other peculiarities in genetics could be subject to revised analysis with the discovery of dynamic mutations. Multifactorial inheritance is usually assumed for phenotypes that "run in families" but fit no simple Mendelian model of inheritance. However, a single gene inherited in a dominant fashion but with penetrance determined by an unstable sequence with the ability to vary into and out of the penetrant range in different generations might appear "multifactorial." In any of the disorders discussed here, it is only the tendency for the repeat sequence to enlarge that allows definition of the genetic segregation and the anticipation phenomenon. In the DM gene, for instance, if the repeat were as capable of reducing in size as it is of expanding, we would find a very complicated clinical picture, with the disease being surely familial, but with difficulty in assigning linkage due to the many nonpenetrant individuals. HD has been termed the "classic" autosomal dominant disorder, and this is apparently due to the very low probability of reduction of the number of repeats into the nonpenetrant range. However, if the mutations in sperm of HD fathers were more balanced, i.e., if it included as many reductions of the same magnitude as increases, there would be many "escapees," individuals with the appropriate region of chromosome 4, but without symptoms (or very late onset). This would add to the already difficult issue of performing genetic linkage studies on a late-onset disorder. Are there other unstable repeats with different properties involved in some of the more common familial (but without simple Mendelian inheritance) maladies? Diabetes, heart disease, high blood pressure, neural tube defects, and cancer-proneness could all be candidates. This picture becomes especially complicated when we consider the possibility of variation limited to somatic tissues, especially if that variation is incomplete and stochastic in early development, with the potential to affect some but not all tissues, with the ensuing complication of the clinical analysis. In this scenario, the lack of a clear clinical syndrome could frustrate any capacity to define the potential genetic basis of the defects.

Imprinting

Imprinting (the expression of genes contributed from one or the other but not both parental chromosomes in a tissue [Nicholls et al. 1989; Hall

1990]) could have a basis in DNA instability. The peculiar parental sex-specific events found in the behavior of the repeats in these disorders suggests that repeat expansion and subsequent methylation with its associated loss of mRNA production might provide a mechanism by which imprints could be transmitted. In this case, the events found in some of these disorders (DM, FRAXA, and FRAXE) could result from improper recognition of signals for normal imprinting (Laird 1987). Similarly, the finding in HD and SCA1 that sperm tend to bear increased repeat numbers over eggs suggests a similarly directed event (Laird 1990).

Directed events?

Finally, it is possible that the events involved in variation of the repeat sequences at these loci are part of a normal developmental system designed to introduce variation into DNA in various tissues. The precedents from the immune system are quite clear; DNA rearranges to create diversity in immunoglobulins and receptors. These events may be the first glimpses of another mechanism for introducing variability into DNA sequence in organ systems or organisms. Since this variation is found at both the germ cell and somatic levels, it is tempting to speculate that these events could be directed by evolutionary pressures. Is this a mechanism for generating species diversity by programed genome-wide changes? Studies of CAG repeats in the involucrin genes of mammals suggest such a mechanism may be at work (Green and Dijan 1992). Other studies of purine-rich trinucleotide repeats suggest an ongoing process on an evolutionary time scale by which "cryptic" repeats can be converted into pure repeats and back again (Gostout et al. 1993). It is likely that selective pressure will be exerted in cases such as the disorders described here against the pure repeats. However, it is not unreasonable to imagine cases of selection for some of these types of trinucleotide mutations, maintaining the machinery involved in generating the variation.

Acknowledgments

The author acknowledges his many friends, colleagues, and collaborators whose insight into the questions raised by this article have helped to shape it. Thanks to Stephen Warren and Julia Parrish for critical reading of the manuscript, to the funding support of the National Institute of Child Health and Human Development (1-RO1-HD29256), and to the Hereditary Disease Foundation for its Dallas workshop on trinucleotides (September 18–19, 1993), which helped in the completion of this manuscript.

References

Aaltonen, L.A., P. Peltomäki, F.S. Leach, P. Sistonen, L. Pylkkänen, J.-P. Mecklin, H. Järvinen, S.M. Powell, J. Jen, S.R. Hamilton, G.M. Petersen, K.W. Kinzler, B. Vogelstein, and A. de la Chapelle. 1993. Clues to the pathogenesis of familial colorectal cancer. *Science* **260**: 812.

Abeliovich, D., I. Lerer, I. Pashut-Lavon, E. Shmueli, A. Raas-Rothschild, and M. Frydman. 1993. Negative expansion of the myotonic dystrophy unstable sequence. *Am. J. Hum. Genet.* **52**: 1175.

Abitbol, M., C. Menini, A.-L. Delezoide, T. Rhyner, M. Vekemans, and J. Mallet. 1993. Nucleus basalis magnocellularis and hippocampus are the major sites of FMR-1 expression in the human fetal brain. *Nature Genet.* **4**: 147.

Andrew, S.E., Y.P. Goldberg, B. Kremer, H. Telenius, J. Theilmann, S. Adam, E. Starr, F. Squitieri, B. Lin, M.A. Kalchman, R.K. Graham, and M.R. Hayden. 1993. The relationship between trinucleotide (CAG) length and clinical features of Huntington's disease. *Nature Genet.* **4**: 398.

Ashley, C.T., Jr., K.D. Wilkinson, D. Reines, and S.T. Warren. 1993a. FMR1 protein contains conserved RNP-family domains and demonstrates selective RNA binding. *Science* **262**: 563.

Ashley, C.T., J.S. Sutcliffe, C.B. Kunst, H.A. Leiner, E.E. Eichler, D.L. Nelson, and S.T. Warren. 1993b. Human and murine FMR-1: Alternative splicing and translational initiation downstream of the CGG-repeat. *Nature Genet.* **4**: 244.

Barceló, J.M., M.S. Mahadevan, C. Tsilfidis, A.E. MacKenzie, and R.G. Korneluk. 1993. Intergenerational stability of the myotonic dystrophy protomutation. *Hum. Mol. Genet.* **2**: 705.

Bell, M.V., M.C. Hirst, Y. Nakahori, R.N. MacKinnon, A. Roche, T.J. Flint, P.A. Jacobs, N. Tommerup, L. Tranebjaerg, U. Froster-Iskenius, B. Kerr, G. Turner, R.H. Lindenbaum, R. Winter, M. Pembrey, S. Thibodeau, and K.E. Davies. 1991. Physical mapping across the fragile X: Hypermethylation and clinical expression of the fragile X syndrome. *Cell* **64**: 861.

Bellen, H.J., H. Vaessin, E. Bier, A. Kolodkin, D. Evelyn, S. Kooyer, and Y.N. Jan. 1992. The *Drosophila couch potato* gene: An essential gene required for normal adult behavior. *Genetics* **131**: 365.

Benson, M. and V. Pirrotta. 1988. The *Drosophila* zeste protein binds cooperatively to sites in many gene regulatory regions: Implications for transvection and gene regulation. *EMBO J.* **7**: 3907.

Biancalana, V., F. Serville, J. Pommier, J. Julien, A. Hanauer, and J.L. Mandel. 1992. Moderate instability of the trinucleotide repeat in spino bulbar muscular atrophy. *Hum. Mol. Genet.* **1**: 255.

Brook, J.D. 1993. Retreat of the triplet? *Nature Genet.* **3**: 279.

Brook, J.D., M.E. McCurrach, H.G. Harley, A.J. Buckler, D. Church, H. Aburatani, K. Hunter, V.P. Stanton, J.P. Thirion, T. Hudson, R. Sohn, B. Zemelman, R.G. Snell, S.A. Rundle, S. Crow, J. Davies, P. Shelbourne, J. Buxton, C. Jones, V. Juvonen, K. Johnson, P.S. Harper, D.J. Shaw, and D.E. Housman. 1992. Molecular basis of myotonic dystrophy: Expansion of a trinucleotide (CTG) repeat at the 3' end of a transcript encoding a protein kinase family member. *Cell* **68**: 799.

Caskey, C.T., A. Pizzuti, Y.H. Fu, R.G. Fenwick, Jr., and D.L. Nelson. 1992. Trip-

let repeat mutations in human disease. *Science* **256**: 784.
Chakravarti, A. 1992. Fragile X founder effect? *Nature Genet.* **1**: 237.
Courey, A.J., D.A. Holtzman, S.P. Jackson, and R. Thian. 1989. Synergistic activation by the glutamine-rich domains of human transcription factor Sp1. *Cell* **59**: 827.
De Boulle, K., A.J.M.H. Verkerk, E. Reyniers, L. Vits, J. Hendrickx, B. Van Roy, F. Van Den Bos, E. de Graaff, B.A. Oostra, and P.J. Willems. 1993. A point mutation in the FMR-1 gene associated with fragile X mental retardation. *Nature Genet.* **3**: 31.
Devys, D., Y. Lutz, N. Rouyer, J.-P. Bellocq, and J.-L. Mandel. 1993. The FMR-1 protein is cytoplasmic, most abundant in neurons and appears normal in carriers of a fragile X premutation. *Nature Genet.* **4**: 335.
Devys, D., V. Biancalana, F. Rousseau, J. Boue, J.L. Mandel, and I. Oberlé. 1992. Analysis of full mutation fragile X mutations in fetal tissues and monozygotic twins indicate that abnormal methylation and somatic heterogeneity are established early in development. *Am. J. Med. Genet.* **43**: 208.
Duyao, M., C. Ambrose, R. Myers, A. Novelletto, F. Persichetti, M. Frontali, S. Folstein, C. Ross, M. Franz, M. Abbott, J. Gray, P. Connealy, A. Young, J. Penney, Z. Hollingsworth, I. Shoulson, A. Lazzarini, A. Falek, W. Koroshetz, D. Sax, E. Bird, J. Vonsattel, E. Bonilla, J. Alvir, J. Bickham Conde, J.-H. Cha, L. Dure, F. Gomez, M. Ramos, J. Sanchez-Ramos, S. Snodgrass, M. de Young, N. Wexler, C. Moscowitz, G. Penchaszadeh, H. MacFarlane, M. Anderson, B. Jenkins, J. Srinidhi, G. Barnes, J. Gusella, and M. MacDonald. 1993. Trinucleotide repeat length instability and age of onset in Huntington's disease. *Nature Genet.* **4**: 387.
Edwards, A., A. Civitello, H.A. Hammond, and C.T. Caskey. 1991. DNA typing and genetic mapping with trimeric and tetrameric tandem repeats. *Am. J. Hum. Genet.* **49**: 746.
Farrer, L.A., L.A. Cupples, D.K. Kiely, P.M. Connealy, and R.H. Myers. 1992. Inverse relationship between age at onset of Huntington disease and paternal age suggests involvement of genetic imprinting. *Am. J. Hum. Genet.* **50**: 528.
Fu, Y.-H., D.L. Friedman, S. Richards, J.A. Pearlman, R.A. Gibbs, A. Pizzuti, T. Ashizawa, M.B. Perryman, G. Scarlato, R.G. Fenwick, Jr., and C.T. Caskey. 1993. Decreased expression of myotonin-protein kinase messenger RNA and protein in adult form of myotonic dystrophy. *Science* **260**: 235.
Fu, Y.H., D.P.A. Kuhl, A. Pizzuti, M. Pieretti, J.S. Sutcliffe, S. Richards, A.J.M.H. Verkerk, J.J.A. Holden, R.G. Fenwick, Jr., S.T. Warren, B.A. Oostra, D.L. Nelson, and C.T. Caskey. 1991. Variation of the CGG repeat at the fragile X site results in genetic instability: Resolution of the Sherman paradox. *Cell* **67**: 1047.
Fu, Y.H., A. Pizzuti, R.G. Fenwick, Jr., J. King, S. Rajnarayan, P.W. Dunne, J. Dubel, G.A. Nasser, T. Ashizawa, P. de Jong, B. Wieringa, R. Korneluk, M.B. Perryman, H.F. Epstein, and C.T. Caskey. 1992. An unstable triplet repeat in a gene related to myotonic muscular dystrophy. *Science* **255**: 1256.
Gedeon, A.K., E. Baker, H. Robinson, M.W. Partington, B. Gross, A. Manca, B. Korn, A. Poustka, S. Yu, G.R. Sutherland, and J.C. Mulley. 1992. Fragile X syndrome without CCG amplification has an FMR1 deletion. *Nature Genet.*

1: 341.
Gibbs, M., A. Collick, R.G. Kelly, and A.J. Jeffreys. 1993. A tetranucleotide repeat mouse minisatellite displaying substantial somatic instability during early preimplantation development. *Genomics* **17:** 121.
Gostout, B., Q. Liu, and S.S. Sommer. 1993. "Cryptic" repeating triplets of purines and pyrimidines (cRRY(i)) are frequent and polymorphic: Analysis of coding cRRY(i) in the proopiomelanocortin (POMC) and TATA-binding protein (TBP) genes. *Am. J. Hum. Genet.* **52:** 1182.
Green, H. 1993. Human genetic diseases due to codon reiteration: Relationship to an evolutionary mechanism. *Cell* **74:** 955.
Green, H. and P. Dijan. 1992. Consecutive actions of different gene-altering mechanisms in the evolution of involucrin. *Mol. Biol. Evol.* **9:** 977.
Hagerman, R.J. 1991. Physical and behavioral phenotype. In *Fragile X syndrome, diagnosis, treatment and research* (ed. R.J. Hagerman and A.C. Silverman), p. 3. Johns Hopkins University Press, Baltimore.
Hall, J.G. 1990. Genomic imprinting: Review and relevance to human diseases. *Am. J. Hum. Genet.* **46:** 857.
Hansen, R.S., S.M. Gartler, C.R. Scott, S.-H. Chen, and C.D. Laird. 1993. Methylation analysis of CGG sites in the CpG island of the human FMR1 gene. *Hum. Mol. Genet.* **1:** 571.
Harley, H.G., J.D. Brook, S.A. Rundle, S. Crow, W. Reardon, A.J. Buckler, P.S. Harper, D.E. Housman, and D.J. Shaw. 1992. Expansion of an unstable DNA region and phenotypic variation in myotonic dystrophy. *Nature* **355:** 545.
Harley, H.G., K.V. Walsh, S. Rundle, J.D. Brook, M. Sarfarazi, M.C. Koch, J.L. Floyd, P.S. Harper, and D.J. Shaw. 1991. Localization of the myotonic dystrophy locus to 19q13.2-19q13.3 and its relationship to twelve polymorphic loci on 19q. *Hum. Genet.* **87:** 73.
Harley, H.G., S.A. Rundle, J.C. MacMillan, J. Myring, J.D. Brook, S. Crow, W. Reardon, I. Fenton, D.J. Shaw, and P.S. Harper. 1993. Size of the unstable CTG repeat sequence in relation to phenotype and parental transmission in myotonic dystrophy. *Am. J. Hum. Genet.* **52:** 1164.
Harper, P.S. 1989. *Myotonic dystrophy*. W.B. Saunders, Philadelphia.
Harper, P.S., H.G. Harley, W. Reardon, and D.J. Shaw. 1992. Anticipation in myotonic dystrophy: New light on an old problem. *Am. J. Hum. Genet.* **51:** 10.
Heitz, D., D. Devys, G. Imbert, C. Kretz, and J.-L. Mandel. 1992. Inheritance of the fragile X syndrome: Size of the fragile X premutation is a major determinant of the transition to full mutation. *J. Med. Genet.* **29:** 794.
Hinds, H.L., C.T. Ashley, J.S. Sutcliffe, D.L. Nelson, S.T. Warren, D.E. Housman, and M. Schalling. 1993. Tissue specific expression of FMR-1 provides evidence for a functional role in fragile X syndrome. *Nature Genet.* **3:** 36.
Hofmann-Radvanyi, H., C. Lavedan, J.-P. Rabès, D. Savoy, C. Duros, K. Johnson, and C. Junien. 1993. Myotonic dystrophy: Absence of CTG enlarged transcript in congenital forms, and low expression of the normal allele. *Hum. Mol. Genet.* **2:** 1263.
Hornstra, I.K., D.L. Nelson, S.T. Warren, and T.P. Yang. 1993. High resolution methylation analysis of the FMR1 gene trinucleotide region in fragile X syndrome. *Hum. Mol. Genet.* **2:** 1659.
Huntington's Disease Collaborative Research Group. 1993. A novel gene contain-

ing a trinucleotide repeat that is expanded and unstable on Huntington's disease chromosomes. *Cell* **72:** 971.

Igarashi, S., Y. Tanno, O. Onodera, M. Yamazaki, S. Sato, A. Ishikawa, N. Miyatani, Y. Ishikawa, K. Sahashi, T. Ibi, T. Miyatake, and S. Tsuji. 1992. Strong correlation between the number of CAG repeats in androgen receptor genes and clinical onset of features of spinal and bulbar muscular atrophy. *Neurology* **42:** 2300.

Imbert, G., C. Kretz, K. Jonhnson, and J.-L. Mandel. 1993. Origin of the expansion mutation in myotonic dystrophy. *Nature Genet.* **4:** 72.

Jansen, G., M. Bartolomei, V. Kalscheuer, G. Merkx, N. Wormskamp, E. Mariman, D. Smeets, H.-H. Ropers, and B. Wieringa. 1993. No imprinting involved in the expression of DM-kinase mRNAs in mouse and human tissues. *Hum. Mol. Genet.* **2:** 1221.

Jansen, G., M. Mahadevan, C. Amemiya, N. Wormskamp, B. Segers, W. Hendriks, K. O'Hoy, S. Baird, L. Sabourin, G. Lennon, P.L. Jap, D. Iles, M. Coerwinkel, M. Hofker, A.V. Carrano, P.J. de Jong, R.G. Korneluk, and B. Wieringa. 1992. Characterization of the myotonic dystrophy region predicts multiple protein isoform-encoding mRNAs. *Nature Genet.* **1:** 261.

Kelly, R., M. Gibbs, A. Collick, and A.J. Jeffreys. 1991. Spontaneous mutation at the hypervariable mouse minisatellite locus Ms6-hm: Flanking DNA sequence and analysis of early and somatic mutation events. *Proc. R. Soc. Lond. B.* **245:** 235.

Kelly, R., G. Bulfield, A. Collick, M. Gibbs, and A.J. Jeffreys. 1989. Characterization of a highly unstable mouse minisatellite locus: Evidence for somatic mutation during early development. *Genomics* **5:** 844.

Knight, S.J.L., A.V. Flannery, M.C. Hirst, L. Campbell, Z. Christodoulou, S.R. Phelps, J. Pointon, H.R. Middleton-Price, A. Barnicoat, M.E. Pembrey, J. Holland, B.A. Oostra, M. Bobrow, and K.E. Davies. 1993. Trinucleotide repeat amplification and hypermethylation of a CpG island in FRAXE mental retardation. *Cell* **74:** 127.

Kremer, E.J., M. Pritchard, M. Lynch, S. Yu, K. Holman, E. Baker, S.T. Warren, D. Schlessinger, G.R. Sutherland, and R.I. Richards. 1991. Mapping of DNA instability at the fragile X to a trinucleotide repeat sequence p(CCG)n. *Science* **252:** 1711.

Laird, C.D. 1987. Proposed mechanism of inheritance and expression of the human fragile-X syndrome of mental retardation. *Genetics* **117:** 587.

———. 1990. Proposed genetic basis of Huntington's disease. *Trends Genet.* **6:** 242.

La Spada, A.R., E.M. Wilson, D.B. Lubahn, A.E. Harding, and K.H. Fischbeck. 1991. Androgen receptor gene mutations in X-linked spinal and bulbar muscular atrophy. *Nature* **352:** 77.

La Spada, A.R., D.B. Roling, A.E. Harding, C.L. Warner, R. Spiegel, I. Hausmanowa-Petrusewicz, W.-C. Yee, and K.H. Fischbeck. 1992. Meiotic stability and genotype-phenotype correlation of the trinucleotide repeat in X-linked spinal and bulbar muscular atrophy. *Nature Genet.* **2:** 301.

MacDonald, M.E., A. Novelletto, C. Lin, D. Tagle, G. Barnes, G. Bates, S. Taylor, B. Allitto, M. Altherr, R. Myers, H. Lehrach, F.S. Collins, J.J. Wasmuth, M. Fronali, and J.F. Gusella. 1992. The Huntington's disease candidate region exhibits many different haplotypes. *Nature Genet.* **1:** 99.

Mahadevan, M., C. Tsilfidis, L. Sabourin, G. Shutler, C. Amemiya, G. Jansen, C.

Neville, M. Narang, J. Barceló, K. O'Hoy, S. Leblond, J. Earle-Macdonald, P.J. de Jong, B. Wieringa, and R.G. Korneluk. 1992. Myotonic dystrophy mutation: An unstable CTG repeat in the 3' untranslated region of the gene. *Science* **255**: 1253.

McConkie-Rosell, A., A.M. Lachiewicz, G.A. Spiridigliozzi, J. Tarleton, S. Schoenwald, M.C. Phelan, P. Goonewardena, X. Ding, and W.T. Brown. 1993. Evidence that methylation of the FMR-1 locus is responsible for variable phenotypic expression of the fragile X syndrome. *Am. J. Hum. Genet.* **53**: 800.

Morton, N.E. and J.N. Macpherson. 1992. Population genetics of the fragile-X syndrome: Multiallelic model for the *FMR1* locus. *Proc. Natl. Acad. Sci.* **89**: 4215.

Nelson, D.L. and S.T. Warren. 1993. Trinucleotide repeat instability: When and where? *Nature Genet.* **4**: 107.

Nicholls, R.D., J.H.M. Knoll, M.G. Butler, S. Karam, and M. Lalande. 1989. Genomic imprinting suggested by maternal heterodisomy in non-deletion Prader-Willi syndrome. *Nature* **342**: 281.

Nussbaum, R.L. and D.H. Ledbetter. 1989. The fragile X syndrome. In *The metabolic basis of inherited disease*, 6th edition (ed. C.R. Scriver et al.), p. 327. McGraw-Hill, New York.

Oberlé, I., F. Rousseau, D. Heitz, C. Kretz, D. Devys, A. Hanauer, J. Boué, M.F. Bertheas, and J.L. Mandel. 1991. Instability of a 550-base pair DNA segment and abnormal methylation in fragile X syndrome. *Science* **252**: 1097.

O'Hoy, K.L., C. Tsilfidis, M.S. Mahadevan, C.E. Neville, J. Barceló, A.G.W. Hunter, and R.G. Korneluk. 1993. Reduction in size of the myotonic dystrophy trinucleotide repeat mutation during transmission. *Science* **259**: 809.

Orr, H.T., M.-Y. Chung, S. Banfi, T.J. Kwiatkowski, Jr., A. Servadio, A.L. Beaudet, A.E. McCall, L.A. Duvick, L.P.W. Ranum, and H.Y. Zoghbi. 1993. Expansion of an unstable trinucleotide (CAG) repeat in spinocerebellar ataxia type 1. *Nature Genet.* **4**: 221.

Oudet, C., E. Mornet, J.L. Serre, F. Thomas, S. Lentes-Zengerling, C. Kretz, C. Deluchat, I. Tejada, J. Boué, A. Boue, and J.L. Mandel. 1993. Linkage disequilibrium between the fragile X mutation and two closely linked CA repeats suggests that fragile X chromosomes are derived from a small number of founder chromosomes. *Am. J. Hum. Genet.* **52**: 297.

Pieretti, M., F. Zhang, Y.H. Fu, S.T. Warren, B.A. Oostra, C.T. Caskey, and D.L. Nelson. 1991. Absence of expression of the *FMR-1* gene in fragile X syndrome. *Cell* **66**: 817.

Redman, J.S., R.G. Fenwick, Jr., Y.-H. Fu, A. Pizzuti, and C.T. Caskey. 1993. Relationship between parental trinucleotide GCT repeat length and severity of myotonic dystrophy in offspring. *J. Am. Med. Assoc.* **269**: 1960.

Reyniers, E., L. Vits, K. De Boulle, B. Van Roy, D. Van Velzen, E. de Graaff, A.J.M.H. Verkerk, H.Z.J. Jorens, J.K. Darby, B.A. Oostra, and P.J. Willems. 1993. The full mutation in the FMR-1 gene of male fragile X patients is absent in their sperm. *Nature Genet.* **4**: 143.

Richards, R.I. and G.R. Sutherland. 1992. Dynamic mutations: A new class of mutations causing human disease. *Cell* **70**: 709.

Richards, R.I., K. Holman, S. Yu, and G.R. Sutherland. 1993. Fragile X syndrome unstable element, p(CCG)n, and other simple tandem repeat sequences are binding sites for specific nuclear proteins. *Hum. Mol. Genet.* **2:** 1429.

Richards, R.I., K. Holman, K. Friend, E. Kremer, D. Hillen, A. Staples, W.T. Brown, P. Goonewardena, J. Tarleton, C. Schwartz, and G.R. Sutherland. 1992. Evidence of founder chromosomes in fragile X syndrome. *Nature Genet.* **1:** 257.

Rousseau, F., D. Heitz, V. Biancalana, S. Blumenfield, C. Kretz, J. Boué, N. Tommerup, C. van der Hagen, C. DeLozier-Blanchet, M.F. Croquette, S. Gilgenkrantz, P. Jalbert, M.A. Voelckel, I. Oberlé, and J.L. Mandel. 1991. Direct diagnosis by DNA analysis of the fragile X syndrome of mental retardation. *N. Engl. J. Med.* **325:** 1673.

Sabouri, L.A., M.S. Mahadevan, M. Narang, D.S.C. Lee, L.C. Surh, and R.G. Korneluk. 1993. Effect of the myotonic dystrophy (DM) mutation on mRNA levels of the DM gene. *Nature Genet.* **4:** 233.

Shelbourne, P., R. Winqvist, E. Kunert, J. Davies, J. Leisti, H. Thiele, H. Bachmann, J. Buxton, B. Williamson, and K. Johnson. 1992. Unstable DNA may be responsible for the incomplete penetrance of the myotonic dystrophy phenotype. *Hum. Mol. Genet.* **1:** 467.

Sherman, S.L., N.E. Morton, P.A. Jacobs, and G. Turner. 1984. The marker (X) syndrome: A cytogenetic and genetic analysis. *Annu. Hum. Genet.* **48:** 21.

Sherman, S.L., P.A. Jacobs, N.E. Morton, U. Froster-Iskenius, P.N. Howard-Peebles, K.B. Nielsen, M.W. Partington, G.R. Sutherland, G. Turner, and M. Watson. 1985. Further segregation analysis of the fragile X syndrome with special reference to transmitting males. *Hum. Genet.* **69:** 289.

Siomi, H., M.C. Siomi, R.L. Nussbaum, and G. Dreyfuss. 1993. The protein product of the fragile X gene, FMR1, has characteristics of an RNA-binding protein. *Cell* **74:** 291.

Snell, R.G., J.C. MacMillan, J.P. Cheadle, I. Fenton, L.P. Lazarou, P. Davies, M.E. MacDonald, J.F. Gusella, P.S. Harper, and D.J. Shaw. 1993. Relationship between trinucleotide repeat expansion and phenotypic variation in Huntington's disease. *Nature Genet.* **4:** 393.

Stern, C. 1973. *Principles of human genetics*, p. 375. W.H. Freeman, San Francisco.

Strand, M., T.A. Prolla, R.M. Liskay, and T.D. Petes. 1993. Destabilization of tracts of simple repetitive DNA in yeast by mutations affecting DNA mismatch repair. *Nature* **365:** 274.

Sutcliffe, J.S., D.L. Nelson, F. Zhang, M. Pieretti, C.T. Caskey, D. Saxe, and S.T. Warren. 1992. DNA methylation represses *FMR-1* transcription in fragile X syndrome. *Hum. Mol. Genet.* **1:** 397.

Sutherland, G.R. and E. Baker. 1992. Characterisation of a new rare fragile site easily confused with the fragile X. *Hum. Mol. Genet.* **1:** 111.

Sutherland, G.R., E.A. Haan, E. Kremer, M. Lynch, M. Pritchard, S. Yu, and R.I. Richards. 1991. Hereditary unstable DNA: A new explanation for some old genetic questions? *Lancet* **338:** 289.

Thibodeau, S.N., G. Bren, and D. Schaid. 1993. Microsatellite instability in cancer of the proximal colon. *Science* **260:** 816.

Tsilfidis, C., A.E. MacKenzie, G. Mettler, J. Barceló, and R.G. Korneluk. 1992. Correlation between CTG trinucleotide repeat length and frequency of severe congenital myotonic dystrophy. *Nature Genet.* **1:** 192.

Vaessin, H., E. Grell, E. Wolff, E. Bier, L.Y. Jan, and Y.N. Jan. 1991. *prospero* is

expressed in neuronal precursors and encodes a nuclear protein that is involved in the control of axonal outgrowth in *Drosophila. Cell* **67:** 941.
Verheij, C., C.E. Bakker, E. de Graaff, J. Keulemans, R. Willemsen, A.J.M.H. Verkerk, H. Galjaard, A.J.J. Reuser, A.T. Hoogeveen, and B.A. Oostra. 1993. Characterization and localization of the FMR-1 gene product associated with the fragile X syndrome. *Nature* **363:** 722.
Verkerk, A.J.M.H., M. Pieretti, J.S. Sutcliffe, Y.H. Fu, D.P.A. Kuhl, A. Pizutti, O. Reiner, S. Richards, M.F. Victoria, F. Zhang, B.E. Eussen, G.J.B. van Ommen, L.A.J. Blonden, G.J. Riggins, J.L. Chastain, C.B. Kunst, H. Galjaard, C.T. Caskey, D.L. Nelson, B.A. Oostra, and S.T. Warren. 1991. Identification of a gene (*FMR-1*) containing a CGG repeat coincident with a breakpoint cluster region exhibiting length variation in fragile X syndrome. *Cell* **65:** 905.
Vincent, A., D. Heitz, C. Petit, C. Kretz, I. Oberlé, and J.L. Mandel. 1991. Abnormal pattern detected in fragile-X patients by pulsed-field gel electrophoresis. *Nature* **349:** 624.
Weber, J.L. and C. Wong. 1993. Mutation of human short tandem repeats. *Hum. Mol. Genet.* **2:** 1123.
Wharton, K.A., K.M. Hohansen, T. Xu, and S. Artavanis-Tsakonas. 1985. Nucleotide sequence from the neurogenic locus notch implies a gene product that shares homology with proteins containing EGF-like repeats. *Cell* **43:** 567.
Willems, P.J., B. Van Roy, K. De Boulle, L. Vits, E. Reyniers, O. Beck, J.E. Dumon, A. Verkerk, and B. Oostra. 1992. Segregation of the fragile X mutation from an affected male to his normal daughter. *Hum. Mol. Genet.* **1:** 511.
Wöhrle, D., I. Hennig, W. Vogel, and P. Steinbach. 1993. Mitotic stability of fragile X mutations in differentiated cells indicates early post-conceptional trinucleotide repeat expansion. *Nature Genet.* **4:** 140.
Wöhrle, D., D. Kotzot, M.C. Hirst, A. Manca, B. Korn, A. Schmidt, G. Barbi, H.D. Rott, A. Poustka, K.E. Davies, and P. Steinbach. 1992. A microdeletion of less than 250 kb, including the proximal part of the FMR-1 gene and the fragile-X site, in a male with the clinical phenotype of fragile-X syndrome. *Am. J. Hum. Genet.* **51:** 299.
Yu, S., J. Mulley, D. Loesch, G. Turner, A. Donnelly, A. Gedeon, D. Hillen, E. Kremer, M. Lynch, M. Pritchard, G.R. Sutherland, and R.I. Richards. 1992. Fragile-X syndrome: Unique genetics of the heritable unstable element. *Am. J. Hum Genet.* **50:** 968.
Yu, S., M. Pritchard, E. Kremer, M. Lynch, J. Nancarrow, E. Baker, K. Holman, J.C. Mulley, S.T. Warren, D. Schlessinger, G.R. Sutherland, and R.I. Richards. 1991. Fragile X genotype characterized by an unstable region of DNA. *Science* **252:** 1179.

Capturing a CAGey Killer

Marcy E. MacDonald,[1,2] Christine M. Ambrose,[1,2]
Mabel P. Duyao,[1,2] and James F. Gusella[1,3]

[1]Molecular Neurogenetics Unit
Massachusetts General Hospital
Charlestown, Massachusetts 02129

[2]Department of Neurology
Harvard Medical School
Boston, Massachusetts 02114

[3]Department of Genetics
Harvard Medical School
Boston, Massachusetts 02114

Huntington's disease (HD) is an inevitably fatal disorder characterized by progressive choreic movements and mental deterioration due to a loss of neurons, chiefly in the striatum. The culprit gene displays dominant inheritance and is located in 4p16.3. For the past decade, investigators have tracked this anonymous killer, gradually closing in on its location. Recently, an expanded, unstable $(CAG)_n$ triplet repeat was identified as the culpable party. The characteristics of the repeat's behavior explain many of the intriguing features of HD inheritance. However, finding the culprit has not yet explained how it accomplishes its ruinous actions.

Main points discussed include:

❑ the search for the HD gene

❑ the identification of the mutation as a triplet repeat

❑ the behavior of the triplet repeat in familial and sporadic cases

❑ relationship to other expanded triplet repeats

❑ impact on diagnosis

❑ possible mechanisms of action

The traces of the offender

HD is an untreatable neurodegenerative disorder caused by a dominant genetic defect located in 4p16.3 (Martin and Gusella 1983; Folstein 1989). The hallmark of HD is a characteristic movement disorder involving progressive chorea. However, the manifestations of the disease gene also include psychiatric and cognitive alterations. The behavioral and motor abnormalities associated with HD are caused by the premature death of neurons, principally in the striatum (Vonsattel et al. 1985). The symptoms of HD can begin at any time in life, but the vast majority of cases occur in middle age. In the youngest victims of the disorder, the course is more rapid and severe. The majority of these early-onset cases result from paternal transmission of the HD defect (Merrit et al. 1969; Bird et al. 1974).

On the most-wanted list for a decade

In 1983, HD was the first disease gene mapped to a chromosomal location using only linkage analysis with polymorphic DNA markers (Gusella et al. 1983, 1984). The mapping of the HD gene to 4p16.3 not only spawned a torrent of similar linkage studies in other inherited disorders, but also initiated a monumental effort to isolate the disease gene on the basis of its location (Gusella et al. 1984; Gusella 1989, 1991). This location cloning strategy passed through a number of stages as the technologies available for human genome analysis evolved rapidly. Somatic cell hybrid mapping panels that dissected 4p16.3 (MacDonald et al. 1987; Smith et al. 1988; Lin et al. 1991) supported the isolation of large numbers of restriction fragment length polymorphism (RFLP) (Gilliam et al. 1987; Wasmuth et al. 1988; Allitto et al. 1991; Lin et al. 1991), variable number of tandem repeat (VNTR) (Wasmuth et al. 1988; MacDonald et al. 1989a), and simple sequence repeat (SSR) markers (Taylor et al. 1992a). These were used to create detailed genetic linkage maps (MacDonald et al. 1989b; Youngman et al. 1989; Allitto et al. 1991) and to anchor the construction of physical maps (Bucan et al. 1990; Bates et al. 1991) and overlapping clone sets (Bates et al. 1992; Baxendale et al. 1993). A progressive refinement of the location for the HD defect was provided by recombination analysis in disease pedigrees (Whaley et al. 1988; MacDonald et al. 1989b; Snell et al. 1992), by linkage disequilibrium between the disorder and individual markers (Snell et al. 1989; Theilmann et al. 1989; MacDonald et al. 1991), and ultimately by the use of multi-allele marker haplotypes to target a small region shared by one third of disease chromosomes (MacDonald et al. 1992). In the end, the detailed inspection of candidate genes (Ambrose et al. 1992; Taylor et al. 1992b; Duyao et al. 1993a) merged with a continued search for informative polymorphisms to reveal a polymorphic $(CAG)_n$ trinucleotide repeat in the 5′ end of a long, novel transcript (Fig. 1)

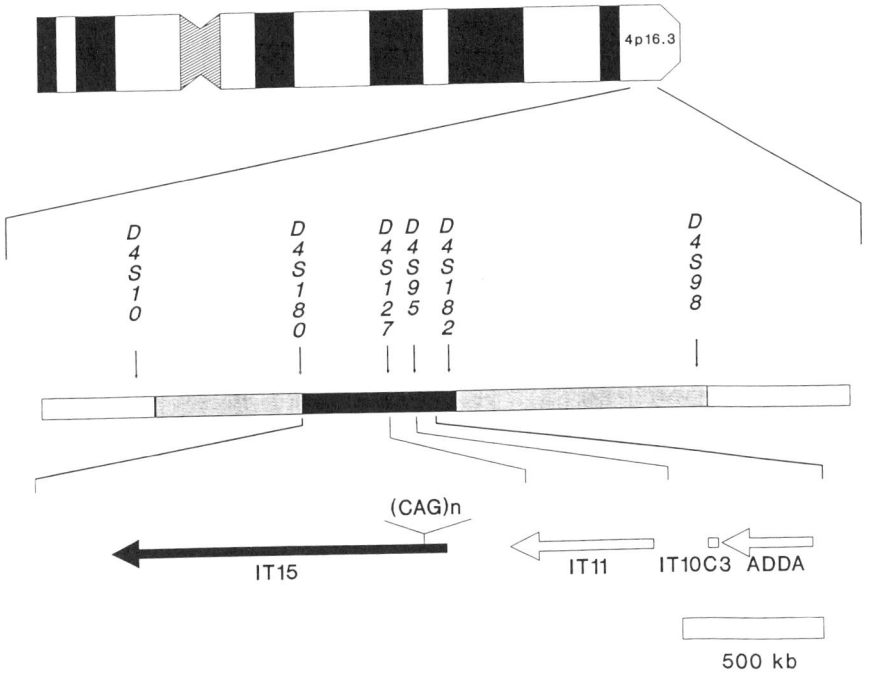

Figure 1 Location of the HD gene in band 4p16.3. A schematic of the chromosome 4 short arm is shown with a portion of 4p16.3 expanded to depict the HD candidate region defined by recombination events (*shaded line*) and the segment in linkage disequilibrium with HD (*filled line*). The latter region is again expanded to show the transcripts that have been identified (Ambrose et al. 1992; Taylor et al. 1992b; Duyao et al. 1993a), including the *IT15* gene containing the unstable (CAG)$_n$ repeat (Huntington's Disease Collaborative Research Group 1993).

(Huntington's Disease Collaborative Research Group 1993). The recognition that this triplet stretch is expanded and unstable on HD chromosomes signaled that the marathon search for the defect was over and that a new quest for understanding of its pathogenesis had begun.

Description of the culprit

Normal chromosomes carry an array of (CAG)$_n$ repeat sizes at the *HD* locus, ranging from 11 to 34 repeat units in length (Duyao et al. 1993b). The distribution of repeat lengths on HD chromosome ranges from 37 to 86 copies. These expanded triplet repeat stretches are very unstable and exhibit fluctuations in length in more than 80% of *HD* transmissions. In contrast to the behavior of the expanded triplet repeat on HD chromosomes, the shorter, polymorphic (CAG)$_n$ stretch found on normal chromosomes is inherited in a stable Mendelian fashion. Figure 2 shows results of polymerase chain reaction (PCR) amplification of normal and HD chromosomes with various repeat lengths.

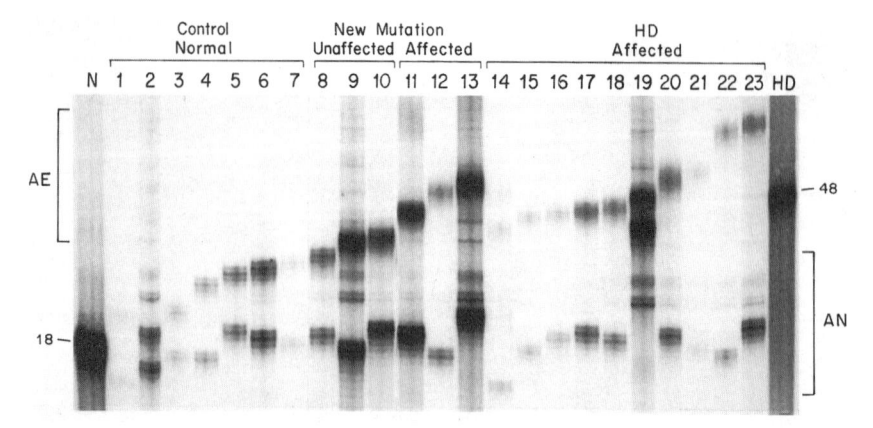

Figure 2 PCR amplification of the HD trinucleotide repeat. The HD triplet repeat was amplified from seven DNAs from normal controls (lanes *1–7*), ten DNAs from affected members of unrelated HD families (lanes *14–23*) and affected (lanes *11–13*) members of three new mutation families and one member from each of these families (lanes *8–10*) who shared the same chromosomal haplotype as the affected new mutation case. Also amplified were two sequenced cosmid controls from a normal chromosome (lane *N*, $(CAG)_{18}$) and from an HD chromosome (lane *HD*, $(CAG)_{48}$). AE encompasses both intermediate range and HD range repeats. AN shows the range of normal allele sizes.

A common *modus operandi*

Haplotype analysis using multi-allele markers has disclosed that only about one third of HD chromosomes appear to be descended from a common ancestor. The large number of different haplotypes which account for the majority of HD chromosomes suggests that there have been multiple independent origins of the HD mutation (MacDonald et al. 1992). Despite the ancestry of the HD chromosome, the same mechanism is operating to cause the disease in all HD families, regardless of ethnic or racial background. An expanded and unstable triplet repeat is found on chromosomes carrying the 26 different haplotypes found in 78 independent families that were studied initially (Huntington's Disease Collaborative Research Group 1993). Our data set has now been enlarged to include chromosomes from 150 families, all of which have an expanded $(CAG)_n$ triplet repeat segregating with the disease (Duyao et al. 1993b).

Reconstructing the criminal's movements

Expansion of the $(CAG)_n$ stretch into the HD range is accompanied by a marked instability of the repeat during meiotic transmission from either mothers or fathers (80% and 84% of transmissions, respectively) (Duyao et al. 1993b). The fluctuations are usually relatively modest in size and can occur as either increases or decreases in triplet repeat length. Transmission from either sex displays a bias toward a size increase rather than a decrease (0.4-unit and 9-unit average increase in female and male transmissions, respectively).

In our data set, neither the magnitude nor the frequency of the size change is apparently affected by the length of the triplet carried on the parental chromosome. The sex of the affected parent does, however, influence the size of the inherited $(CAG)_n$ repeat. Offspring of affected mothers inherited triplet repeat stretches that varied within a narrow range of 8 repeat units surrounding the maternal length. The majority (62%) of offspring of affected fathers inherited repeat units that fell within a 10-unit range surrounding the paternal repeat length, but 38% of progeny of affected fathers inherited $(CAG)_n$ repeats that were much larger (9–42-unit jumps) than the father's allele. The overall distribution of HD repeat lengths reflects the propensity of male transmission to generate larger alleles (Fig. 3). Of the HD chromosomes with repeat lengths greater than 55, 89% were transmitted from the father.

The conspicuous jumps in triplet repeat length during transmission from affected fathers appear to occur during spermatogenesis (Duyao et

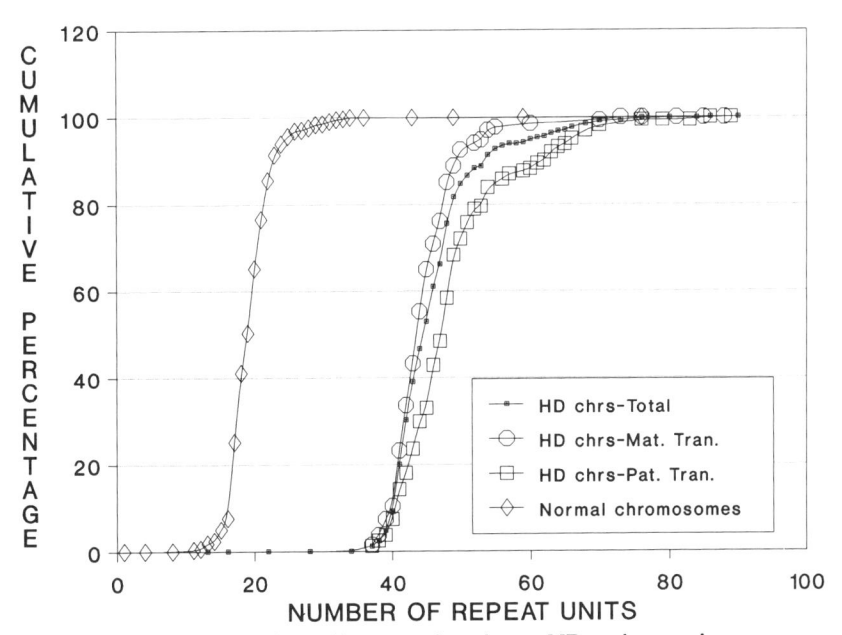

Figure 3 Distribution of $(CAG)_n$ repeat lengths on HD and normal chromosomes. The cumulative frequency of different repeat lengths is shown as a percentage of 425 HD chromosomes (*small squares*) and 545 normal chromosomes (*diamonds*). For those HD chromosomes where the sex of the transmitting parent was known, repeat lengths derived from maternal (*circles*) and paternal (*large squares*) transmissions, 134 and 161 chromosomes, respectively, are plotted separately.

al. 1993b). This is most dramatically demonstrated by the finding that the distribution of $(CAG)_n$ repeat lengths seen in the general HD population can be reflected in the sperm DNA of a single HD victim. In a preliminary screen of HD and control tissues, including lymphoblastoid, blood leukocyte, brain, and peripheral organ DNAs, only HD sperm DNA showed a diffuse pattern of PCR products representing a broad range of expanded triplet repeat sizes on the HD chromosome. These results raise the possibility that the instability in repeat length occurs during the mitotic or meiotic divisions that give rise to the haploid male germ cell. It is possible, but as yet untested, that a similar degree of instability is encountered during female gametogenesis, but that effects of repeat expansion on viability preclude the generation of eggs with vastly greater lengths of $(CAG)_n$. In fact, the influence of repeat expansion on gamete viability in either sex remains to be examined.

Description of the crime

There is a strong statistical association between the length of the triplet repeat and the age of onset of HD (Duyao et al. 1993b). In general, the longer the stretch of $(CAG)_n$, the earlier the disorder manifests, with the very longest repeats causing juvenile-onset HD. The correlation of repeat length with age of onset and the propensity for large increases in allele size to occur during male transmission, therefore, mutually explain the observation that cases of juvenile HD arise most often from affected fathers (Merrit et al. 1969; Bird et al. 1974).

The age of onset in individuals homozygous for the HD mutation (Wexler et al. 1987) illustrates the dominance of the disorder at the molecular level. Each affected parent contributes an expanded $(CAG)_n$ repeat of a different size, and the onset of the disease in the homozygous child matches that predicted for a single expanded allele rather than the sum of the triplet repeat units carried by both HD chromosomes (Huntington's Disease Collaborative Research Group 1993). The mutation, therefore, is completely dominant and exerts the same effect in the context of a normal allele or another expanded $(CAG)_n$ triplet repeat.

Although the correlation between HD repeat length and age of onset is highly significant, each repeat length is, in fact, associated with a broad range in age of onset (Duyao et al. 1993b). For example, the majority (88%) of all HD gene carriers have between 37 and 52 repeat units, and this range is associated with an age of onset that fluctuates by ±18 years around the predicted mean. It is clear from the breadth of this distribution that genetic modifiers, environmental effects, or stochastic processes also affect the precise age at which symptoms begin. As observed in the HD homozygotes, the length of the $(CAG)_n$ repeat on the normal chromosome carried by HD heterozygous individuals also does not influence the age of onset of the disease.

Guilty on most counts

The task of identifying legitimate sporadic cases of HD has proven to be a challenging one (Huntington's Disease Collaborative Group 1993; Myers et al. 1993). This is in part because the frequency of new mutation to HD is very low (~1.8 x 10^{-7} to 4.5 x 10^{-7}) and in part because confirmation of the HD status in the afflicted individuals and their antecedents is made problematic by the relatively late onset of the disorder. Despite these difficulties, unequivocal cases of sporadic HD have been found. These HD victims appear in well-documented families that are free of the disease, they display characteristic symptoms of HD, and some have also transmitted the disease to an affected offspring. In several cases, it has also been possible to determine using DNA marker analysis that other members of the family who show no signs of HD bear the same chromosome 4 homolog as the putative new mutant.

In almost all cases, the presence of the disease in the sporadic affected individual is due to an expanded $(CAG)_n$ stretch that falls within the range of repeat lengths observed on HD chromosomes from typical HD families. The homologous chromosomes borne by the unaffected relatives of these victims, however, display a variety of smaller repeat lengths (34–38 units) in a range overlapping with the tail ends of both the normal (11–34 units) and HD (37–86 units) repeat distributions. In this initial survey, one exceptional family was found in which both the victim and siblings carry a 33-unit triplet repeat.

At present, these intermediate-sized $(CAG)_n$ repeats defy simple classification (Myers et al. 1993). The instability of alleles in the 33–38-unit range has not yet been explored. There is no evidence that these alleles are more mutable than are alleles in the normal range. Consequently, they could represent rare high normal alleles that are usually inherited in a stable Mendelian fashion. Alternatively, the distribution underlying $(CAG)_n$ length may be continuous, and triplet repeats of 33–38 units could represent unstable HD alleles that are associated with extremely late onset of the disease. In this segment of the continuum, other genetic factors, environmental influences, or stochastic processes may play a prominent role in determining whether the disorder occurs. A third possibility is that the intermediate alleles represent a novel premutation state which is clinically innocuous but inclined to increased instability.

Resolving the status of repeat lengths that do not fall squarely within either the normal or the HD size ranges will require an extensive evaluation of both the instability of these intermediate length $(CAG)_n$ stretches and their frequency in the normal population. Moreover, postmortem brain tissue from individuals carrying these alleles will have to be examined to determine the neuropathological impact of intermediate-sized triplet repeats.

A criminal predisposition?

The major 4p16.3 haplotype, shared by 1/3 of all HD families, is also present in the majority of families in which sporadic HD arises (Myers et al. 1993). This provocative finding suggests that the major haplotype represents a predisposing haplotype which provides a reservoir of chromosomes that independently undergo mutation to HD. However, the instability of $(CAG)_n$ stretches on this haplotype has not been evaluated, and the finding might be explained by the association of a relatively large normal allele with this particular haplotype.

Revisiting the crime scene

The $(CAG)_n$ repeat is located within a large gene, *IT15* (for *i*nteresting *t*ranscript 15), that is defined by a set of overlapping cDNA clones (Fig. 1) (Huntington's Disease Collaborative Research Group 1993). The composite sequence of *IT15* predicts a long open reading frame that could encode a 348-kD protein which bears no significant extended similarity to previously identified proteins. Knowledge of the primary sequence of the HD gene product, therefore, provides no clues as to the normal function of the protein which has been dubbed *huntingtin*. The presence of an ATG, methionine, initiator codon upstream of the $(CAG)_n$ stretch suggests, but does not prove, that the HD mutation results in the expansion of a polymorphic stretch of polyglutamines at the amino terminus of the protein. Alternatively, translation conceivably could initiate at a position downstream from this start site, putting the HD defect in the 5′-untranslated region of the gene.

Other members of the gang

HD is the fourth in a growing series of unrelated genetic disorders whose underlying defect is the expansion of a trinucleotide repeat. Although HD may share a common mutational mechanism with each of these three disorders, the precise details of repeat expansion are different in each case. Fragile X syndrome is caused by expansion of a $(GGC)_n$ trinucleotide repeat (Fu et al. 1991; Kremer et al. 1991; Pieretti et al. 1991; Verkerk et al. 1991; Richards et al. 1992; Yu et al. 1992; Oudet et al. 1993) that is associated with inactivation of the *FMR-1* gene. The triplet is located in the 5′-untranslated region of *FMR-1*, and the clinical manifestations of the disease are associated with spectacular size increases (100s or 1000s of units) which most commonly occur during transmission from the mother. New mutations arise from an intermediate, predisposing class of repeat which is larger than that found on normal chromosomes and is associated with limited, size-dependent instability.

The $(CTG)_n$ (CAG on the reverse strand) repeat whose expansion causes myotonic dystrophy is found in the 3' end of the myotonin protein kinase gene (Brook et al. 1992; Fu et al. 1992; Harley et al. 1992a,b; Mahadevan et al. 1992; Suthers et al. 1992; Tsilfidis et al. 1992). The mechanism of action of this repeat is not known, but it probably acts at the level of the mRNA. Like fragile X syndrome, phenotypic consequences are associated with prodigious repeat lengths comprising hundreds or thousands of units. New mutations involving large expansions that occur in transmission from either sex occur on a predisposing chromosome (Harley et al. 1991).

Spino-bulbar muscular atrophy, like HD, is caused by the expansion of a $(CAG)_n$ repeat (LaSpada et al. 1991, 1992; Biancalana et al. 1992). The triplet, located within the coding sequence of the androgen receptor, increases in size, producing an altered protein with a variable polyglutamine stretch. The distributions of the triplet repeat lengths in normal and disease chromosomes are similar to that seen in HD, and there is a tendency for increases in length to occur during male transmission. However, specific features of the $(CAG)_n$ variation in the two disorders are different, most noticeably in the more restricted instability of the repeat in spino-bulbar muscular atrophy. The repeat length is more stable during transmission, and when alterations do occur, the magnitude of the change is modest, in contrast to the relatively large increases that can occur during male transmission in HD.

Despite the shared *modus operandi*, it is apparent that the behavior and the repercussions of triplet repeat expansion are quite different for the three disorders that have been studied in greatest detail. Thus, although it is obviously necessary to examine the pathogenesis of each disorder separately, the possibility of different mechanisms of instability must also be evaluated.

How did he do it?

It is not self-evident how expansion of $(CAG)_n$ in the context of the *IT15* gene leads to selective loss of striatal neurons, but several possibilities can be tested. At the moment, the only mechanism that can be excluded is that of transcriptional inactivation (Huntington's Disease Collaborative Research Group 1993; C.M. Ambrose et al., in prep.). The *IT15* gene is expressed as a pair of mRNAs (about 10 kb and 13 kb, depending on the length of the 3' exon) in a wide variety of tissues, including the brain, where the effects of the expansion are most prominent. HD homozygotes express both species of *IT15* mRNA despite the presence of sometimes very long stretches of $(CAG)_n$, suggesting that repeat expansion does not eliminate *IT15* gene expression. Although these experiments have been done using lymphoblastoid cell lines, it is likely that normal levels of mRNA will also be detected in HD brain in the regions most affected by the repeat. Moreover, patients with Wolf-Hirschhorn syndrome who carry heterozygous deletions of 4p16.3 often have only one copy of the *IT15* gene and do not display any features of HD.

The triplet repeat expansion may act at the level of the protein. If the repeat is translated so that the amino terminus of *huntingtin* contains a variable number of glutamine residues, the complete dominance of the disease phenotype could arise because the mutation endows the protein with a new property. For example, elongation of the polyglutamine stretch may allow the protein to interact with a novel cellular component, to incapacitate the normal protein or to change the intracellular location of the protein. The $(CAG)_n$ expansion may act at the level of the mRNA by virtue of its location either in the protein or in the 5'-untranslated portion of the gene. Repeat expansion may alter the efficiency of translation, the half-life, the splicing pattern, or other cellular processes impacting on or involving participation of the mRNA. It remains possible that the triplet repeat expansion acts at the level of the DNA to alter the expression of a neighboring gene. However, with the exception of this last possibility, all hypotheses must account for the specific loss of striatal neurons caused by a gene which appears to be ubiquitously expressed.

Fingerprints on record

Discovery of triplet repeat expansion as the cause of the disorder has profound consequences for diagnosis in HD. The most immediate impact is that individuals can be directly diagnosed using a PCR amplification assay of the triplet repeat. This will provide a more widely applicable, more rapid, and less expensive alternative to predictive testing by linkage analysis. Presymptomatic and prenatal testing using linked DNA markers (Meissen et al. 1988; Brandt et al. 1989) is unwieldy and can be applied only to "at risk" individuals for whom several family members are willing and able to donate their DNA. The ability to examine the triplet repeat directly in cases suspected to be HD but without a corroborating family history will also be of considerable interest to the HD community.

Determination of HD gene carrier status using the PCR assay to determine triplet repeat length also has its difficulties. The major impediment is the inability to easily classify repeats that overlap the tail-ends of the distributions of the normal and HD repeat lengths. Currently, among members of well-documented HD pedigrees, only three repeat units separate a chromosome that can cause HD and one that does not (Duyao et al. 1993b). Larger data sets could well eliminate this gap. A vivid example of the problems this can create for diagnosis is provided by the intermediate-sized repeat lengths carried in families in which sporadic HD has occurred (Myers et al. 1993). Unaffected family members carry repeats in the 34-38 repeat unit range, overlapping the tails of the normal and HD distributions. In one exceptional case, 33 repeat units is associated with the probable presence of HD. It is, therefore, not appropriate to assign a clinical outcome for an individual who carries an allele size that falls in the 34-38 repeat unit range. Moreover, for these individuals it is also not possible to predict the risk of transmission of HD to their children because the frequency of expansion and instability of repeats in this size range is not known.

Not guilty on a few counts?

Another confounding issue is raised by the identification of a few individuals within large HD pedigrees who were diagnosed with HD but who appear to have inherited a normal chromosome based on linked markers (MacDonald et al. 1989b; Pritchard et al. 1992). All other affected members of these families possess an expanded $(CAG)_n$ repeat. Although gene conversion or double crossover had been suggested as a likely possibility for the source of the nonconforming individuals, all of them in fact inherited the normal range triplet repeat characteristic of the affected parent's normal chromosome defined by marker analysis. Although it is possible that these individuals have been misdiagnosed due to the combination of their possessing a motor abnormality and also being members of an established HD pedigree, they have not yet been explained satisfactorily. Regardless of their ultimate cause, the occurrence of such individuals at a low but measurable frequency must be considered as an additional source of confusion in presymptomatic testing of at-risk individuals.

Counseling the victims

The delivery of information to individuals carrying a gene defect with psychiatric consequences is fraught with danger. With the advent of linkage testing, programs that include intensive genetic counseling have been established (Meissen et al. 1988; Brandt et al. 1989). The nature of the information provided by the direct PCR test is not intrinsically different from that delivered in a linkage test, and consequently, the existing protocols should be adhered to after minor modifications to deal with the specific technical features of the PCR test.

The length of the $(CAG)_n$ repeat will in most cases provide unequivocal information about a person's gene carrier status. However, the length of the repeat unit should not be used to impart information about the age of onset of the disease. Any given repeat can be associated with a relatively broad range of ages of onset, making such predictions meaningless for any single individual.

The sentence for the detectives

Discovery of the HD gene defect is a first step on the path to understanding the pathogenesis of the disorder. It is now necessary to define the second, and subsequent, steps along the pathway from the trinucleotide repeat to the ultimate loss of neurons in the striatum. The availability of the defective gene will facilitate the development of genetic animal models of HD that can be used to manipulate and explore the genesis of neuronal cell loss. Each aspect of the biochemistry and pathology of HD that can be discerned could potentially inspire the development of an effective treatment, offering hope to the members of families that suffer this tragic disorder.

Acknowledgments

The authors' work on Huntington's disease is supported by National Institutes of Health grants NS-16367 (Huntington's Disease Center Without Walls) and NS-22031 and by grants from Bristol-Myers Squibb Inc. and the Hereditary Disease Foundation Collaborative Research Agreement. M.P.D. and C.M.A. have been supported by fellowships from the Huntington's Disease Society of America and the Andrew B. Cogan Fellowship of the Hereditary Disease Foundation.

References

Allitto, B.A., M.E. MacDonald, M. Bucan, J. Richards, D. Romano, L.W. Whaley, B. Falcone, J. Ianazzi, N.S. Wexler, J.J. Wasmuth, F.S. Collins, H. Lehrach, J.L. Haines, and J.F. Gusella. 1991. Increased recombination adjacent to the Huntington's disease-linked *D4S10* marker. *Genomics* 9: 104.

Ambrose, C., M. James, G. Barnes, C. Lin, G. Bates, M. Altherr, M. Duyao, N. Groot, D. Church, J.J. Wasmuth, H. Lehrach, D. Housman, A. Buckler, J.F. Gusella, and M.E. MacDonald. 1992. A novel G protein-coupled receptor kinase cloned from 4p16.3. *Hum. Mol. Genet.* 1: 697.

Bates, G.P., M.E. MacDonald, S. Baxendale, S. Youngman, C. Lin, W.L. Whaley, J.J. Wasmuth, J.F. Gusella, and H. Lehrach. 1991. Defined physical limits of the Huntington disease gene candidate region. *Am. J. Hum. Genet.* 49: 7.

Bates, G.P., J. Valdes, H. Hummerich, S. Baxendale, D.L. Le Paslier, A.P. Monaco, D. Tagle, M.E. MacDonald, M. Altherr, M. Ross, B.H. Brownstein, D. Bentley, J.J. Wasmuth, J.F. Gusella, D. Cohen, F. Collins, and H. Lehrach. 1992. Characterization of a yeast artificial chromosome contig spanning the Huntington Disease Gene Candidate Region. *Nature Genet.* 1: 180.

Baxendale, S., M.E. MacDonald, R. Mott, F. Francis, C. Lin, S.F. Kirby, M. James, G. Zehetner, H. Hummerich, J. Valdes, F.S. Collins, J.F. Gusella, H.

Lehrach, and G.P. Bates. 1993. Construction of cosmid contigs and high resolution restriction maps of a 2 Mbp region containing the Huntington's disease gene. *Nature Genet.* **4**: 181.

Biancalana, V., F. Serville, J. Pommier, J. Julien, A. Hanauer, and J.L. Mandel. 1992. Moderate instability of the trinucleotide repeat in spino-bulbar muscular atrophy. *Hum. Mol. Genet.* **1**: 255.

Bird, E.D., A.J. Caro, and J.B. Pilling. 1974. A sex related factor in the inheritance of Huntington's chorea. *Ann. Hum. Genet.* **37**: 255.

Brandt, J., K.A. Quaid, S.E. Folstein, P. Garber, N.E. Maestri, M.H. Abbott, P.R. Slavney, M.L. Franz, L. Kasch, and H.H. Kazazian. 1989. Presymptomatic diagnosis of delayed-onset disease with linked DNA markers: The experience in Huntington's disease. *J. Am. Med. Assoc.* **261**: 3108.

Brook, J.D., M.E. McCurrach, H.G. Harley, A.J. Buckler, D. Church, H. Aburatani, K. Hunter, V.P. Stanton, J.P. Thirion, T. Hudson, R. Sohn, B. Zemelman, R.G. Snell, S.A. Rundle, S. Crow, J. Davies, P. Shelbourne, J. Buxton, C. Jones, V. Juxonen, K. Johnson, P.S. Harper, D.J. Shaw, and D.E. Housman. 1992. Molecular basis of myotonic dystrophy: Expansion of a trinucleotide (CTG) repeat at the 3' end of a transcript encoding a protein kinase family member. *Cell* **68**: 799.

Bucan, M., M. Zimmer, W.L. Whaley, A. Poustka, S. Youngman, B.A. Allitto, E. Ormondroyd, B. Smith, T.M. Pohl, M. MacDonald, G. Bates, J. Richards, S. Volinia, T.C. Gilliam, Z. Sedlacek, F.S. Collins, J.J. Wasmuth, D.J. Shaw, J.F. Gusella, A.M. Frischauf, and H. Lehrach. 1990. Physical maps of 4p16.3, the area expected to contain the Huntington's disease mutation. *Genomics* **6**: 1.

Duyao, M.P., S.A.M. Taylor, A.J. Buckler, C.M. Ambrose, C. Lin, N. Groot, D. Church, G. Barnes, J.J. Wasmuth, D.E. Housman, M.E. MacDonald, and J.F. Gusella. 1993a. A gene from the Huntington's disease candidate region with similarity to a superfamily of transporter proteins. *Hum. Mol. Genet.* (in press).

Duyao, M.P., C.M. Ambrose, R.H. Myers, A. Novelletto, F. Persichetti, M. Frontali, S.E. Folstein, C. Ross, M.L. Franz, M. Abbott, J. Gray, P.M. Conneally, A. Young, J. Penney, Z. Hollingsworth, I. Shoulson, A.M. Lazzarini, A. Falek, W. Koroshetz, D.S. Sax, E. Bird, J.P. Vonsattel, E. Bonilla, J. Alvir, J. Bickham Conde, J.-H. Cha, L. Dure, F. Gomez, M. Ramos, J. Sanchez-Ramos, S.R. Snodgrass, M. de Young, N.S. Wexler, C. Moscowitz, G. Penchaszadeh, H. MacFarlane, M.A. Anderson, B. Jenkins, J. Srinidhi, G. Barnes, J.F. Gusella, and M.E. MacDonald. 1993b. Trinucleotide repeat length instability and age of onset in Huntington's Disease. *Nature Genet.* (in press).

Folstein, S.E. 1989. *Huntington's disease. A disorder of families*. The Johns Hopkins University Press, Baltimore.

Fu, Y.H., D.P.A. Kuhl, A. Pizzuti, M. Pieretti, J.S. Sutcliffe, S. Richards, A.J.M.H. Verkerk, J. Holden, R.J. Fenwick, S.T. Warren, B.A. Oostra, D.L. Nelson, and C.T. Caskey. 1991. Variation of the CGG repeat at the fragile X site results in genetic instability: resolution of the Sherman paradox. *Cell* **67**: 1047.

Fu, Y.H., A. Pizzuti, R.G. Fenwick, J. King, Jr., S. Rajnarayan, P.W. Dunne, J. Dubel, G.A. Nasser, T. Ashizawa, P. DeJong, B. Wieringa, R. Korneluk, M.B. Perryman, H.F. Epstein, and C.T. Caskey. 1992. An unstable triplet repeat in a gene related to myotonic muscular dystrophy. *Science* **255**: 1256.

Gilliam, T.C., R.E. Tanzi, J.L. Haines, T.I. Bonner, A.G. Faryniarz, W.J. Hobbs, M.E. MacDonald, S.V. Cheng, S.E. Folstein, P.M. Conneally, N.S. Wexler, and J.F. Gusella. 1987. Localization of the Huntington's disease gene to a small segment of chromosome 4 flanked by *D4S10* and the telomere. *Cell* **50**: 565.

Gusella, J.F. 1989. Location cloning strategy for characterizing genetic defects in Huntington's disease and Alzheimer's disease. *FASEB J.* **3**: 2036.

———. 1991. Huntington's disease. *Adv. Hum. Genet.* **20**: 125.

Gusella, J.F., R.E. Tanzi, M.A. Anderson, W. Hobbs, K. Gibbons, R. Raschtchian, T.C. Gilliam, M.R. Wallace, N.S. Wexler, and P.M. Conneally. 1984. DNA markers for nervous system diseases. *Science* **225**: 1320.

Gusella, J.F., N.S. Wexler, P.M. Conneally, S.L. Naylor, M.A. Anderson, R.E. Tanzi, P.C. Watkins, K. Ottina, M.R. Wallace, A.Y. Sakaguchi, A.B. Young, I. Shoulson, E. Bonilla, and J.B. Martin. 1983. A polymorphic DNA marker genetically linked to Huntington's Disease. *Nature* **306**: 234.

Harley, H.G., J.D. Brook, S.A. Rundle, S. Crow, W. Reardon, A.J. Buckler, P.S. Harper, and D.E. Housman. 1992a. Expansion of an unstable DNA region and phenotypic variation in myotonic dystrophy. *Nature* **355**: 545.

Harley, H.G., S.A. Rundle, W. Reardon, J. Myring, S. Crow, J.D. Brook, P.S. Harper, and D.J. Shaw. 1992b. Unstable DNA sequence in myotonic dystrophy. *Lancet* **339**: 1125.

Harley, H.G., J.D. Brook, J. Floyd, S.A. Rundle, S. Crow, K.V. Walsh, M.C. Thibault, P.S. Harper, and D.J. Shaw. 1991. Detection of linkage disequilibrium between the myotonic dystrophy locus and a new polymorphic DNA marker. *Am. J. Hum. Genet.* **49**: 68.

Huntington's Disease Collaborative Research Group. 1993. A novel gene containing a trinucleotide repeat that is expanded and unstable on Huntington's disease chromosomes. *Cell* **72**: 971.

Kremer, E.J., M. Pritchard, M. Lynch, S. Yu, K. Holman, E. Baker, S.T. Warren, D. Schlessinger, G.R. Sutherland, and R.I. Richards. 1991. Mapping of DNA instability at the fragile X to a trinucleotide repeat sequence p(CCG)n. *Science* **252**: 1711.

LaSpada, A.R., E.M. Wilson, D.B. Lubahn, A.E. Harding, and H. Fishbeck. 1991. Androgen receptor gene mutations in X-linked spinal and bulbar muscular atrophy. *Nature* **352**: 77.

LaSpada, A.R., D.B. Roling, A.E. Harding, C.L. Warner, R. Speigel, I. Hausmanowa-Petrusewicz, W.-C. Yee, and K.H. Fischbeck. 1992. Meiotic stability and genotype-phenotype correlation of the trinucleotide repeat in X-linked spinal and bulbar muscular atrophy. *Nature Genet.* **2**: 301.

Lin, C.S., M. Altherr, G. Bates, W.L. Whaley, A.P. Read, R. Harris, H. Lehrach, J.J. Wasmuth, J.F. Gusella, and M.E. MacDonald. 1991. New DNA markers in the Huntington's disease gene candidate region. *Somat. Cell Mol. Genet.* **17**: 481.

MacDonald, M.E., C. Lin, L. Srinidhi, G. Bates, M. Altherr, W.L. Whaley, H. Lehrach, J. Wasmuth, and J.F. Gusella. 1991. Complex patterns of linkage disequilibrium in the Huntington disease region. *Am. J. Hum. Genet.* **49**: 723.

MacDonald, M.E., M.A. Anderson, T.C. Gilliam, L. Tranebjerg, N.J. Carpenter, E.

Magenis, M.R. Hayden, S.T. Healey, T.I. Bonner, and J.F. Gusella. 1987. A somatic cell hybrid panel for localizing DNA segments near the Huntington's disease gene. *Genomics* **1:** 29.

MacDonald, M.E., S.V. Cheng, M. Zimmer, J.L. Haines, A.M. Poustka, B.A. Allitto, B. Smith, W.L. Whaley, D. Romano, J. Jagadeesh, H. Lehrach, J.J. Wasmuth, A.M. Frischauf, and J.F. Gusella. 1989a. Clustering of multiallele DNA markers near the Huntington's disease gene. *J. Clin. Invest.* **84:** 1013.

MacDonald, M.E., A. Novelletto, C. Lin, D. Tagle, G. Barnes, G. Bates, S. Taylor, B. Allitto, M. Altherr, R. Myers, H. Lehrach, F.S. Collins, J.J. Wasmuth, M. Frontali, and J.F. Gusella. 1992. The Huntington's disease candidate region exhibits many different haplotypes. *Nature Genet.* **1:** 99.

MacDonald, M.E., J.L. Haines, M. Zimmer, S.V. Cheng, S. Youngman, W.L. Whaley, M. Bucan, A.B. Allitto, B. Smith, J. Leavitt, A.M. Poustka, P. Harper, H. Lehrach, J.J. Wasmuth, A.M. Frischauf, and J.F. Gusella. 1989b. Recombination events suggest possible locations for the Huntington's disease gene. *Neuron* **3:** 183.

Mahadevan, M., C. Tsilfidis, L. Sabourin, G. Shutler, C. Amemiya, G. Jansen, C. Neville, M. Narang, J. Barcelo, K. O'Hoy, S. Leblond, J. Earle-MacDonald, P.J. DeJong, B. Wieringa, and G. Korneluk. 1992. Myotonic dystrophy mutation: An unstable CTG repeat in the 3′ untranslated region of the gene. *Science* **255:** 1253.

Martin, J.B. and J.F. Gusella. 1986. Huntington's disease: Pathogenesis and management. *N. Engl. J. Med.* **315:** 1267.

Meissen, G.J., R.H. Myers, C.A. Mastromauro, W.J. Koroshetz, K.W. Klinger, L.A. Farrer, P.A. Watkins, J.F. Gusella, E.D. Bird, and J.B. Martin. 1988. Predictive testing for Huntington's disease with use of a linked DNA marker. *N. Engl. J. Med.* **318:** 535.

Merrit, A.D., P.M. Conneally, N.F. Rahman, and A.L. Drew. 1969. Juvenile Huntington's chorea. In *Progress in neurogenetics* (ed. A. Barbeau and J.R. Brunette), p. 645. Excerpta Medica, Amsterdam.

Myers, R.H., J. Leavitt, L.A. Farrer, J. Jagadeesh, H. McFarlane, R.J. Mark, and J.F. Gusella. 1989. Homozygote for Huntington's disease. *Am. J. Hum. Genet.* **45:** 615.

Myers, R.H., M.E. MacDonald, W.J. Koroshetz, M.P. Duya, C.A. Ambrose, S.A.M. Taylor, G. Barnes, J. Srinidhi, C.S. Lin, W.L. Whaley, A.M. Lazzarini, M. Schwarz, G. Wolff, E.D. Bird, J.-P. Vonsattel, and J.F. Gusella. 1993. De novo expansion of a $(CAG)_n$ repeat in sporadic Huntington's disease. *Nature Genet.* (in press).

Oudet, C., E. Mornet, J.L. Serre, F. Thomas, S. Lentes-Zengerling, C. Kretz, C. Deluchat, I. Tejada, J. Boue, A. Boue, and J.L. Mandel. 1993. Linkage disequilibrium between the fragile X mutation and two closely linked CA repeats suggests that fragile X chromosomes are derived from a small number of founder chromosomes. *Am. J. Hum. Genet.* **52:** 297.

Pieretti, M., F. Zhang, Y.H. Fu, S.T. Warren, B.A. Oostra, C.T. Caskey, and D.L. Nelson. 1991. Absence of expression of the *FMR-1* gene in fragile X syndrome. *Cell* **66:** 817.

Pritchard, C., N. Zhu, J. Zuo, L. Bull, M.A. Pericak-Vance, A.D. Roses, A. Milatovich, U. Francke, D.R. Cox, and R.M. Myers. 1992. Recombination of 4p16 DNA markers in an unusual family with Huntington disease. *Am.*

J. Hum. Genet. **50:** 1218.
Richards, R.I., K. Holman, K. Friend, E. Kremer, D. Hillen, A. Staples, W.T. Brown, P. Goonewardena, J. Tarleton, C. Schwartz, and G.R. Sutherland. 1992. Fragile X syndrome: evidence of founder chromosomes. *Nature Genet.* **1:** 257.
Smith, B., D. Skarecky, U. Bengtsson, R.E. Magenis, N. Carpenter, and J.J. Wasmuth. 1988. Isolation of DNA markers in the direction of the Huntington disease gene from the G8 locus. *Am. J. Hum. Genet.* **42:** 335.
Snell, R.G., L. Lazarou, S. Youngman, O.W.J. Quarrell, J.J. Wasmuth, D.L. Shaw, and P.S. Harper. 1989. Linkage disequilibrium in Huntington's disease: An improved localization for the gene. *J. Med. Genet.* **26:** 673.
Snell, R.G., L.M. Thompson, D.A. Tagle, T.L. Holloway, G. Barnes, H.G. Harley, L.A. Sandkuijl, M.E. MacDonald, F.S. Collins, J.F. Gusella, P.S. Harper, and D.J. Shaw. 1992. A recombination event that redefines the Huntington disease region. *Am. J. Hum. Genet.* **51:** 357.
Suthers, G.K., S.M. Huson, and K.E. Davies. 1992. Instability versus predictability: The molecular diagnosis of myotonic dystrophy. *J. Med. Genet.* **29:** 761.
Taylor, S.A.M., G.T. Barnes, M.E. MacDonald, and J.F. Gusella. 1992a. A dinucleotide repeat polymorphism at the *D4S127* locus. *Hum. Mol. Genet.* **1:** 147.
Taylor, S.A.M., R.G. Snell, A. Buckler, C. Ambrose, M. Duyao, D. Church, C.S. Lin, M. Altherr, G.P. Bates, N. Groot, G. Barnes, D.J. Shaw, H. Lehrach, J.J. Wasmuth, P.S. Harper, D.E. Housman, M.E. MacDonald, and J.F. Gusella. 1992b. Cloning of the α-adducin gene from the Huntington's disease candidate region of chromosome 4 by exon amplification. *Nature Genet.* **2:** 223.
Theilman, J., S. Kanani, R. Shiang, C. Robbins, O. Quarrell, M. Huggins, A. Hedrick, and M.R. Hayden. 1989. Non-random association between alleles detected at *D4S95* and *D4S98* and the Huntington's disease gene. *J. Med. Genet.* **26:** 676.
Tsilfidis, C., A.E. McKenzie, G. Mettler, J. Barcelo, and R.G. Korneluk. 1992. Correlation between CTG trinucleotide repeat length and frequency of severe congential myotonic dystrophy. *Nature Genet.* **1:** 192.
Verkerk, A.J.M.H., M. Pieretti, J.S. Sutcliffe, Y.H. Fu, D.P.A. Kuhl, A. Pizzuti, O. Reiner, S. Richards, M.F. Victoria, R. Zhang, B.E. Eussen, G.B. van Ommen, L. Blonden, G.J. Riggins, J.L. Chastain, C.B. Kunst, H. Galjaard, C.T. Caskey, D.L. Nelson, B.A. Oostra, and S.T. Warren. 1991. Identification of a gene (*FMR-1*) containing a CGG repeat coincident with a breakpoint cluster region exhibiting length variation in fragile X syndrome. *Cell* **65:** 904.
Vonsattel, J.P., R.H. Myers, T.J. Stevens, R.J. Ferrante, E.D. Bird, and E.P. Richardson, Jr. 1985. Neuropathological classification of Huntington's disease. *J. Neuropathol. Exp. Neurol.* **44:** 549.
Wasmuth, J.J., J. Hewitt, B. Smith, G. Allard, J.L. Haines, D. Skarecky, E. Partlow, and M.R. Hayden. 1988. A highly polymorphic locus very tightly linked to the Huntington's disease. *Nature* **332:** 734.
Wexler, N.S., A.B. Young, R.E. Tanzi, H. Travers, S. Starosta-Rubenstein, J.B. Penney, S.R. Snodgrass, I. Shoulson, F. Gomez, M.A. Ramos-Arroyo, G. Penchaszadeh, R. Moreno, K. Gibbons, A. Faryniarz, W. Hobbs, M.A.

Anderson, E. Bonilla, P.M. Conneally, and J.F. Gusella. 1987. Homozygotes for Huntington's disease. *Nature* **326:** 194.

Whaley, W.L., F. Michiels, M.E. MacDonald, D. Rommano, M. Zimmer, B. Smith, J. Leavitt, M. Bucan, J.L. Haines, T.C. Gilliam, G. Zehetner, C. Smith, C.R. Cantor, A.-M. Frischauf, J.J. Wasmuth, H. Lehrach, and J.F. Gusella. 1988. Mapping of *D4S98/S114/S113* confines the Huntington's defect to a reduced physical region at the telomere of chromosome 4. *Nucleic Acids Res.* **16:** 11769.

Youngman, S., M. Sarafafazi, M. Bucan, M.E. MacDonald, B. Smith, M. Zimmer, C. Gilliam, A.M. Frischauf, J.J. Wasmuth, J.F. Gusella, H. Lehrach, P.S. Harper, and D.J. Shaw. 1989. A new DNA marker [*D4S90*] is terminally located on the short arm of chromosome 4 close to the Huntington's disease gene. *Genomics* **5:** 802.

Yu, S., J. Mulley, D. Loesch, G. Turner, A. Donnelly, A. Gedeon, D. Hillen, E. Kremer, M. Lynch, M. Pritchard, G.R. Sutherland, and R.I. Richards. 1992. Fragile-X syndrome: Unique genetics of the heritable unstable element. *Am. J. Hum. Genet.* **50:** 968.

Mechanisms of Mutations at Human Minisatellite Loci

John A.L. Armour, Darren G. Monckton, David L. Neil,
Keiji Tamaki, Annette MacLeod, Maxine Allen,
Moira Crosier, and Alec J. Jeffreys

Department of Genetics
University of Leicester
Leicester LE1 7RH, United Kingdom

Many tandemly repeated minisatellite (VNTR) loci show extremely high levels of variability in human populations. This hypervariability is the result of a high rate of spontaneous germ-line mutation to new length alleles. Initial work analyzing the linkage phase of markers flanking loci undergoing mutation showed that simple unequal recombination between alleles was unlikely to be a common mechanism for these length-change mutations. More recent studies have used the pattern of interspersion of different repeat unit variants within a minisatellite array to explore internal allelic structure. These studies have shown that there is a polarity of variation and mutation at these loci; most of the observed variation in allelic structure and the rearrangements seen in spontaneous mutations are clustered at one extremity of the locus. About 50% of germ-line mutations can be demonstrated to be due to unequal exchange between alleles, but consistent with the earlier studies, these interallelic mutations appear predominantly to involve only short patches of exchange, with no example of simple unequal recombination observed to date.

The main points are:

❏ Allelic diversity at hypervariable minisatellites is generated by a high rate of de novo mutation to new length alleles in the germ line.

❏ Studies of variant repeat interspersion within minisatellite arrays can be used to investigate mutational mechanisms, both by defining internal allelic variability in populations and by the analysis of structural changes occurring during de novo length-change mutations.

❏ The variability and mutation at three human minisatellite loci studied in detail are highly polar, with nearly all repeat unit turnover occurring at or near one end of the locus.

❏ Some de novo mutations occur via unequal exchanges of repeat units between alleles. Where flanking polymorphisms are informative, they have not been exchanged, suggesting that only a short stretch of repeat units is involved in a gene-conversion-like process.

INTRODUCTION

Tandemly repeated loci are highly abundant in the human genome. There is a wide spectrum of repeat sizes (Fig. 1), from the small blocks of reiterated mononucleotides, frequently poly(dA) associated with retroposon tails (Economou et al. 1990), to blocks of centromeric major satellite, of which individual arrays can be as large as 5 Mb (Willard 1991). Figure 1 shows the major size classes of tandemly repeated DNA in the human genome: Dinucleotide repeats (A) have been extensively characterized, mostly as informative markers in segregation analysis (Weber 1990; Weissenbach et al. 1992). Arrays of short repeats up to a total size of about 1 kb (Fig. 1B) have only lately been studied in detail but are the subject of much recent work, both as general genetic markers (Edwards et al. 1991; Riggins et al. 1992) and in the pathogenesis of some inherited disorders (Caskey et al. 1992; Davies 1992). Length variation has also been demonstrated in blocks of major satellite repeats (Fig. 1E) (Willard 1991) and in arrays of 40-bp repeats at the "midisatellite" locus D1Z2 (Fig. 1D) (Nakamura et al. 1987a). It is at the minisatellite loci (Armour and Jeffreys 1992), at which array size generally falls within the range 1-30 kb (C), that our knowledge of mechanisms of the length-change mutations ultimately responsible for length polymorphism is most detailed and with which this review is concerned.

The extent of our knowledge of repeat unit turnover mechanisms at minisatellite loci owes much to their size, which allowed cloning in bacteriophage vectors and analysis by Southern blot hybridization (Jeffreys et al. 1985; Wong et al. 1986, 1987; Nakamura et al. 1987b). The smaller microsatellites became amenable to detailed analysis only after amplification by the polymerase chain reaction (PCR) was possible (Litt and Luty 1989; Tautz 1989; Weber and May 1989), and pulsed field gel electrophoresis was required to resolve the sizes of arrays present at major satellite blocks (Mahtani and Willard 1990; Oakey and Tyler-Smith 1990). The mechanisms of mutation at tandem arrays other than

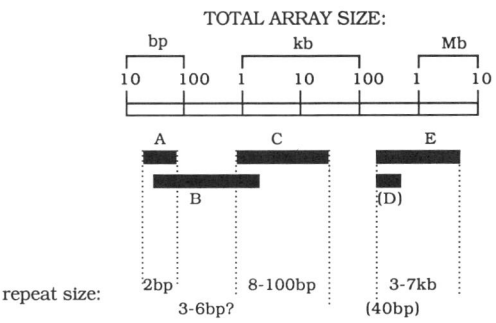

Figure 1 Characterized size ranges of human repeated DNA, including dinucleotide repeats (A), major satellites (E), and minisatellites (C).

minisatellite loci have thus not been studied in the same detail, and the results obtained at minisatellites may serve as a useful initial paradigm for studies of the generation of polymorphism at other tandemly repeated loci.

THE VARIETY OF HUMAN MINISATELLITES

The number of human minisatellite loci has been estimated at approximately 1500 per haploid genome (Braman et al. 1985; Jeffreys 1987). This estimate is based on the number of highly polymorphic loci (PIC >0.7) recovered from a large-scale screen of bacteriophage λ clones of human DNA for polymorphism (Schumm et al. 1985). Although it is a reasonable supposition that loci with this level of informativeness are tandemly repeated in structure, minisatellite loci are clearly underrepresented in standard genomic libraries (Wyman et al. 1985; Kelly et al. 1989; Armour et al. 1990), suggesting that this figure may be an underestimate. Furthermore, this estimate is concerned only with highly polymorphic minisatellites, neglecting entirely an apparently considerable complement of monomorphic minisatellites in the human genome (Armour et al. 1990). Because they have little to offer for genetic analysis, monomorphic repeat arrays have not been studied in detail; nevertheless, they do show that our view of the minisatellite compartment of the genome is subject to considerable bias toward the highly polymorphic and the easily isolated loci.

Analysis of genomic location has shown clearly that there is a strong, but not exclusive, tendency for highly polymorphic minisatellites to localize to the subtelomeric regions of human chromosomes (Royle et al. 1988; Armour et al. 1989b; Vergnaud et al. 1991). Indeed, those subtelomeric regions that have been well characterized, including 16p (Jarman and Wells 1989) and the pseudoautosomal X-Y pairing region (Rouyer et al. 1986; Page et al. 1987), show very high densities of minisatellite loci. Close clustering of minisatellites in subtelomeric regions can sometimes result in the fortuitous isolation of two different minisatellite arrays on a single short cloned DNA fragment (Royle et al. 1988; Armour et al. 1989b; Armour and Jeffreys 1991; Vergnaud et al. 1991).

This subtelomeric localization of minisatellite sequences correlates with regions of high chiasma density in meiosis (Hulten 1974; Laurie and Hulten 1985; Chandley and Mitchell 1988). The linkage maps in these regions also appear to be subject to sex-specific recombination frequencies; many such regions show enhanced recombination in male meiosis and consequent expansion of these regions in the male linkage map (Nakamura et al. 1988, 1989).

Although our current impression may be subject to cloning bias, the autosomes all appear to have similar numbers of polymorphic minisatellites. The X chromosome, in contrast, appears to have relatively few polymorphic minisatellites in its sex-specific region (Donis-Keller et al. 1987; Fraser et al. 1989; Armour et al. 1990; Consalez et al. 1991), whereas the pseudoautosomal X-Y pairing region is extremely rich in polymorphic minisatellites (Cooke et al. 1985; Page et al. 1987). The X-Y pairing region is a part of the X chromosome that has a homolog in male meiosis, perhaps suggesting a specific role for the male germ line in minisatellite evolution.

MINISATELLITE VARIATION AND LENGTH-CHANGE MUTATION

Some minisatellite loci show very high variability, adopting large numbers of different length allelic states and displaying high levels of heterozygosity in human populations (Wong et al. 1986, 1987; Nakamura et al. 1987b; Armour et al. 1990; Vergnaud et al. 1991). The different allelic states contain different numbers of repeat units, and hence are of different overall length. In the absence of selective pressure, such high levels of population variability can be generated by high de novo mutation rates to new length alleles in the germ line. Indeed, at some of these loci, the rates of germ-line mutation to new length alleles are high enough (up to 15% per gamete) to measure directly by pedigree analysis (Jeffreys et al. 1988; Vergnaud et al. 1991). In one study of five highly unstable loci, the rates of length-change mutation at the minisatellites studied bore the relation to their observed heterozygosities predicted by a neutral mutation/random drift model for their evolution (Jeffreys et al. 1988). More recently, smaller tandem repeat arrays, including dinucleotide repeats, have been shown to undergo occasional spontaneous germ-line mutation to new length alleles (Kwiatkowski et al. 1992; Weissenbach et al. 1992). In addition to estimating germ-line mutation rates for minisatellite loci, pedigree analysis can reveal general patterns of mutation: For example, the initial series of loci reported (Jeffreys et al. 1988) appeared to undergo mutation equally in male and female germ lines, and increases and decreases in allele size occurred with approximately equal frequencies. More recent evidence suggests a small but significant male bias (Jeffreys et al. 1991a). At least one locus, however, is much more unstable in the male germ line than in the female germ line and appears to be biased toward increases in allele size (Vergnaud et al. 1991).

A number of mechanisms may be proposed to account for length-change mutations occurring in the germ line, including replication slippage, unequal sister chromatid exchange, and unequal recombination between alleles. Of these, unequal recombination received most initial experimental attention, both because of speculation that minisatellite loci might act as hot spots for recombination, and also because this mechanism would be expected to lead to a characteristic associated change, namely, exchange of flanking markers. A single example of germ-line mutation at the D17S5 locus (YNZ22) was shown not to involve exchange of immediately flanking markers (Wolff et al. 1988). Similarly, a survey of germ-line mutations at the D1S7 (MS1) locus showed that flanking markers within the surrounding 10-cM interval were not exchanged at significantly elevated rates in gametes bearing length-change mutations (Wolff et al. 1989).

In addition to these direct studies of de novo mutation at minisatellite loci, indirect evidence can be obtained by studying allelic disequilibrium at closely flanking markers; any hot spot for recombination would be predicted to act to promote relatively free association between neighboring regions of strong disequilibrium, as observed at the β-globin locus (Antonarakis et al. 1982). These studies have failed to demonstrate recombinant haplotypes accompanying minisatellite diversification and rather suggest that evolution of minisatellite alleles appears to proceed entirely within haplotypic frameworks (Higgs et al. 1986; Cox et al. 1988; Kasperczyk et al. 1990). These studies seem to have excluded simple unequal recombination between alleles as the major mechanism for length-change mutation at these loci; thus, minisatellite mutations do not appear frequently to be accompanied by long-range crossovers, such as would appear in linkage mapping as meiotic recombinations. It should be noted, however, that these studies have limited resolving power; no analysis was made of repeat unit variants, and flanking markers analyzed are frequently many centimorgans away from the locus undergoing mutation. It thus remains possible that minisatellite mutations may frequently entail more involved and especially more *local* recombinational mechanisms (see below).

INTERNAL STRUCTURE AT MINISATELLITE REPEAT LOCI

Almost all minisatellites studied in detail show a second level of variation (Jeffreys et al. 1990, 1991b). In addition to variation in repeat unit copy number, the precise sequence of the tandemly repeated unit can show variation within the array. In most cases, this repeat unit sequence variation is subtle, consisting, for example, of a single variable position in the repeat unit sequence. Nevertheless, even such subtle variation can be exploited to demonstrate the existence of different repeat unit types in the array and to map their pattern of interspersion along an allele, thereby acting as a probe of the internal structure of the tandem array (Jeffreys et al. 1990, 1991b). If the repeat unit sequence variation creates or destroys a site for a restriction enzyme, then restriction mapping can be used to map the disposition of repeat units along an array (Jeffreys et al. 1990). Thus in Figure 2, whereas enzyme E2 cuts every repeat in the array, enzyme E1 cuts some repeats but not others, thus showing the positions within the repeat array of two repeat unit types, containing or lacking the site for E1. More conveniently—and more generally, since it obviates the need for a suitable restriction enzyme—the sequence variation can be sampled directly by using PCR primers specific to each sequence variant, such as P1 and P2 in Figure 2 (Jeffreys et al. 1991b).

Minisatellite variant repeat mapping (MVR mapping) can be used to clarify mutational mechanisms by two general approaches: (1) indirectly, by mapping and comparing alleles from unrelated individuals, allowing the inference of general principles and patterns in the mutation process; (2) by the direct analysis of progenitor and new mutant structures in spontaneous germ-line mutation events, by which specific mutational mechanisms may be inferred. MVR mapping has been applied in our laboratory to the detailed analysis of variation and germ-line mutation at three hypervariable minisatellite loci: MS32 (D1S8) (Jeffreys et al. 1990, 1991b; Tamaki et al. 1992), MS31 (D7S21) (D.L. Neil and A.J. Jeffreys, in prep.), and MS205 (D16S309) (Royle et al. 1992; J.A.L. Armour et al., in prep.). The results obtained suggest that general principles are operating at all three loci; these are discussed below.

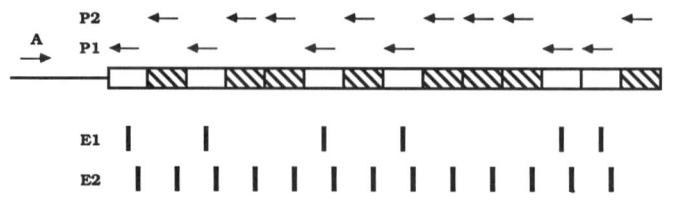

Figure 2 Mapping internal repeat variants within a minisatellite array, using either restriction enzymes (E1 and E2) or PCR primers (P1 and P2).

POLARITY: A MUTATIONAL HOT SPOT

Analysis of allelic variation by MVR mapping has shown clearly that at all three loci studied in detail there is a polarity of variation: Whereas one relatively invariant end of each locus shows only a relatively limited repertoire of variation, the other end is "ultravariable," with virtually no two ends shared between different alleles (Jeffreys et al. 1990, 1991b; Tamaki et al., 1992; D.L. Neil and A.J. Jeffreys; J.A.L. Armour et al.; both in prep.). Moreover, groups of apparently closely related alleles can be found, which differ by small changes at the ultravariable end (Fig. 3). This pattern of variation suggests that during the turnover of repeat units leading to structures seen in modern human populations, rearrangements are preferentially occurring at one end of the array.

It is somewhat surprising, therefore, that only a small region at one end of these minisatellite loci appears to be frequently involved in the generation of new alleles. This finding predicts that de novo mutations involve repeat unit rearrangement predominantly at this end of the locus. Analysis of germ-line mutations observed in pedigrees has so far confirmed this prediction, often with mutational activity confined to the first one or two repeat units at the ultravariable end.

GERM-LINE MUTATION: UNEQUAL EXCHANGES BETWEEN ALLELES

MVR mapping can also be used in the direct analysis of structural changes occurring during de novo length-change mutations observed in pedigrees. Although a relatively large survey at D1S8 has failed to detect structural rearrangements that do not alter allele length (Jeffreys et al. 1991b), it remains possible that such "isometric" rearrangements are contributing to the evolution of minisatellite loci; in practice, observation of length changes remains the most convenient method for ascertainment of mutations at minisatellite loci. The mutation events analyzed so far show, as predicted, a strong polarity toward the ultravariable end of all three loci examined. Some mutations appear to be occurring via entirely intra-allelic mechanisms; others, however, can only be satisfactorily explained as unequal exchanges of repeat units between alleles (Jeffreys et al. 1991b). Examples of events involving exchange of repeat units between alleles, which account for approximately half of all length-change mutations, have now been seen at all three loci studied, suggesting that the process may be a general one (A.J. Jeffreys et al., unpubl.).

Detailed analysis of mutations involving unequal exchanges can elucidate whether the mutation has occurred by a simple unequal recombination, including exchange of flanking markers, or whether the

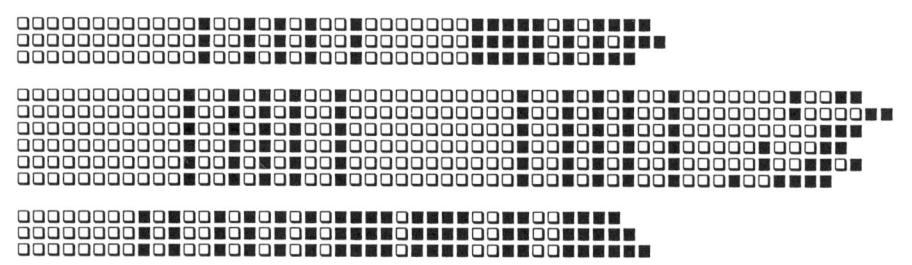

Figure 3 Groups of very similar, presumably closely related, alleles at D16S309 (MS205); alleles differ by small changes at the "ultravariable" end of the array, here shown on the right.

exchange is confined to the repeats. Where flanking polymorphic markers are informative, they have not been exchanged during this kind of mutation (Fig. 4A), suggesting that the event normally involves a small "patch" of exchange between the alleles. Further circumstantial evidence for the involvement of recombinational processes comes from the frequent appearance at the junction between the contributions of donor and recipient of a short region (shaded in Fig. 4A) of anomalous repeats, not corresponding to the cognate position in either the donor or recipient alleles. These anomalous regions of unattributable repeats may reflect the involvement of mismatch repair in the generation of the mutant structure, but in some cases, they are complex and have no obvious origin. Furthermore, some examples of "subterminal" exchanges of

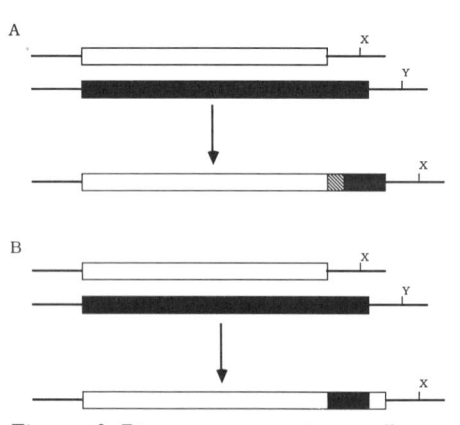

Figure 4 Diagram representing small patches of unequal exchange between alleles, illustrating anomalous (*shaded*) repeats near the exchange junction (A), and "subterminal" insertion of repeats from the donor allele (B). In neither case are alleles (X and Y) at an immediately flanking polymorphism exchanged.

repeat units (Fig. 4B) support the view of interallelic exchange as frequently involving only short regions of the donor allele, inserted into a highly active region in the recipient.

Although this direct evidence from de novo mutation events suggests that interallelic exchange predominantly involves processes akin to gene conversion, is minisatellite mutation ever accompanied by a "full" recombination event, of a kind that would be detected in linkage analysis? The mutation rate at some minisatellite loci is so high that even if only a very small proportion of mutations were due to "simple" unequal recombination, this would represent a greatly enhanced local rate for meiotic recombination, possibly sufficient to account for the generally increased rates of recombination in subtelomeric regions. Although not a single example of such a mutation has been documented to date, some groups of apparently closely related alleles (Fig. 3) occasionally show different flanking haplotypes for nearby substitutional polymorphisms, suggesting that the domain of exchange may occasionally extend beyond the repeat array, well into the flanking single-copy sequence.

On the other hand, these "switches" of flanking haplotypes may simply be the results of larger patches of unequal conversion, rather than true unequal recombinations. If so, then the bias toward conversion rather than recombination might produce anomalous local genetic maps: Thus, whereas two markers on either side of the active domain may show very little recombination, a third marker within the domain may have high frequencies of exchange with both of the flanking markers. In these, the local genetic map may undergo extreme distortion, but on a very small scale.

FUTURE PROSPECTS: MUTATIONAL LOAD

The detailed analysis of minisatellite evolution outlined above derives from studies of allelic structure in human populations and from studies of changes in structure occurring during de novo mutations. Even at the relatively high rates found at minisatellite loci, the incidence of naturally occurring mutations does not allow a number of important questions relating to mutational load to be addressed: For example, How is the rate of mutation to new length alleles related to allele length? What flanking DNA sequences are important in determining instability and polarity? Can variation in instability be found either between alleles or between individuals? These and other questions are not addressed by the study of sporadic mutations detected from family studies; too many potentially important variables differ between different cases.

We are therefore beginning to develop methods for the quantitative recovery of mutant alleles directly from sperm DNA. This approach uses PCR from small pools of sperm DNA, so that the products amplified from individual mutant molecules constitute a detectable fraction of the total signal from each pool; by this approach, many thousand gametes from the same individual can be screened in a single experiment. Thus, not only can germ-line mutations be recovered and analyzed without the need for pedigree analysis, but also the frequency and types of mutation can be assessed for each individual. We may therefore begin, at least in the male germ line, to assess the relative importance of different factors, as outlined above, in the mutational load at minisatellite loci.

This general approach can also be applied to somatic DNA. Minisatellite loci are known to undergo somatic mutation, at high frequency in gastrointestinal carcinomas (Armour et al. 1989a), but also in normal tissues, including blood (Jeffreys et al. 1990). Methods using small-pool PCR to assess mutational load may also be valuable in the analysis of factors important in what now appear to be different processes involved in somatic mutation and may, for example, be applied in cultured cells in the assessment of environmental agents active in the production of minisatellite mutations. Work in progress in our laboratory on human minisatellite sequences introduced into mice may also be instrumental in elucidating factors affecting minisatellite instability.

RELEVANCE TO SHORTER TANDEM REPEATS?

The analysis of mutational mechanisms at minisatellite loci outlined above has provided detailed insights into their evolutionary turnover and has included some surprising findings, including the polarity of mutation and the clear demonstration of recombinational mechanisms rather than slippage as the major mutational process. Are any of these findings relevant to the generation of variation at shorter tandemly repeated loci, including dinucleotide arrays, and the triplet repeat loci involved in human genetic disorders? Examples of de novo mutation at dinucleotide repeats have been observed directly and have not been accompanied by the exchange of (relatively distant) flanking markers (Kwiatkowski et al. 1992; Weissenbach et al. 1992). Similarly, studies of linkage disequilibrium around dinucleotide arrays suggest that, at least on a large scale, most allelic diversification occurs on fixed haplotypic frameworks (Morral et al. 1991; Sherrington et al. 1991). Similarly, de novo mutations at the fragile-X and myotonic dystrophy loci have not been accompanied by exchange of flanking markers (Fu et al. 1991; Harley et al. 1992; Richards et al. 1992; Shelbourne et al. 1992; Yu et al. 1992).

Recombinational mutations at minisatellite loci, however, appear to leave little or no evidence of the exchanges even in very close flanking markers. It may thus be premature to conclude from the evidence outlined above that shorter arrays do not mutate by unequal exchange in the germ line. Indeed, the demonstration of a complex event, involving at least two small patches of interallelic exchange, near the myotonic dystrophy repeats suggests that recombinational mechanisms may yet be involved in length-change mutations at shorter repeat arrays (O'Hoy et al. 1993).

Acknowledgments

J.A.L.A. is a Wellcome Trust postdoctoral fellow. The research of A.J.J. was supported in part by an International Research Scholars award from the Howard Hughes Medical Institute. This work was also supported by grants to A.J.J. from the MRC and the Royal Society.

References

Antonarakis, S.E., C.D. Boehm, P.J.V. Giardina, and H.H. Kazazian. 1982. Nonrandom association of polymorphic restriction sites in the β-globin gene cluster. *Proc. Natl. Acad. Sci.* **79:** 137.

Armour, J.A.L. and A.J. Jeffreys. 1991. STS for minisatellite MS607 (D22S163). *Nucleic Acids Res.* **19:** 3158.

———. 1992. Biology and applications of human minisatellite loci. *Curr. Opin. Genet. Dev.* **2:** 850.

Armour, J.A.L., S. Povey, S. Jeremiah, and A.J. Jeffreys. 1990. Systematic cloning of human minisatellites from ordered array Charomid libraries. *Genomics* **8:** 501.

Armour, J.A.L., I. Patel, S.L. Thein, M. Fey, and A.J. Jeffreys. 1989a. Analysis of somatic mutations at human minisatellite loci in tumours and cell lines. *Genomics* **4:** 328.

Armour, J.A.L., Z. Wong, V. Wilson, N.J. Royle, and A.J. Jeffreys. 1989b. Sequences flanking the repeat arrays of human minisatellites: Association with tandem and dispersed repeat elements. *Nucleic Acids Res.* **17:** 4925.

Braman, J., D. Barker, J. Schumm, R. Knowlton, and H. Donis-Keller. 1985. Characterization of very highly polymorphic RFLP probes. *Cytogenet. Cell Genet.* **40:** 589.

Caskey, C.T., A. Pizzuti, Y.-H. Fu, R.G. Fenwick, Jr., and D.L. Nelson. 1992. Triplet repeat mutations in human disease. *Science* **256:** 784.

Chandley, A.C. and A.R. Mitchell. 1988. Hypervariable minisatellite regions are sites for crossing-over at meiosis in man. *Cytogenet. Cell. Genet.* **48:** 152.

Consalez, G.G., N.S.T. Thomas, C.L. Stayton, S.J.L. Knight, M. Johnson, L.C. Hopkins, P.S. Harper, L.J. Elsas, and S.T. Warren. 1991. Assignment of Emery-Dreyfuss muscular dystrophy to the distal region of Xq28—The results of a collaborative study. *Am. J. Hum. Genet.* **48:** 468.

Cooke, H.J., W.R.A. Brown, and G.A. Rappold. 1985. Hypervariable telomeric sequences from the human sex chromosomes are pseudoautosomal. *Nature* **317:** 687.

Cox, N.J., G.I. Bell, and K. Xiang. 1988. Linkage disequilibrium in the human insulin/insulin-like growth factor II region of human chromosome 11. *Am. J. Hum. Genet.* **43:** 495.

Davies, K.E. 1992. The costs of instability. *Nature* **356:** 15.

Donis-Keller, H., P. Green, C. Helms, S. Cartinhour, B. Weiffenbach, K. Stephens, D.W. Bowden, D.R. Smith, E.S. Lander, D. Botstein, G. Akots, K.S. Rediker, T. Gravius, V.A. Brown, M.B. Rising, C. Parker, J.A. Powers, D.E. Watt, E.R. Kaufman, A. Bricker, P. Phipps, H. Muller-Kahle, T.R. Fulton, S. Ng, J.W. Schumm, J.C. Braman, R.G. Knowlton, D.F. Barker, S.M. Crooks, S.E. Lincoln, M.J. Daly, and J. Abrahamson. 1987. A genetic linkage map of the human genome. *Cell* **51:** 319.

Economou, E.P., A.W. Bergen, A.C. Warren, and S.E. Antonarakis. 1990. The polydeoxyadenylate tract of Alu repetitive elements is polymorphic in the human genome. *Proc. Natl. Acad. Sci.* **87:** 2951.

Edwards, A., A. Civitello, H.A. Hammond, and C.T. Caskey. 1991. DNA typing and genetic mapping with trimeric and tetrameric tandem repeats. *Am. J. Hum. Genet.* **49:** 746.

Fraser, N.J., Y. Boyd, and I. Craig. 1989. Isolation and characterization of a human variable copy number tandem repeat at Xcen-p11.22. *Genomics* **5:** 144.

Fu, Y.-H., D.P.A. Kuhl, A. Pizzuti, M. Pieretti, J.S. Sutcliffe, S. Richards, A.J.M.H. Verkerk. J.J.A. Holden, R.G. Fenwick, Jr., S.T. Warren, B.A. Oostra, D.L. Nelson, and C.T. Caskey. 1991. Variation of the CGG repeat at the fragile X site results in genetic instability: Resolution of the Sherman paradox. *Cell* **67:** 1047.

Harley, H.G., J.D. Brook, S.A. Rundle, S. Crow, W. Reardon, A.J. Buckler, P.S. Harper, D.E. Housman, and D.J. Shaw. 1992. Expansion of an unstable DNA region and phenotypic variation in myotonic dystrophy. *Nature* **355:** 545.

Higgs, D.R., J.S. Wainscoat, J. Flint, A.V.S. Hill, S.L. Thein, R.D. Nicholls, H. Teal, H. Ayyub, T.E.A. Peto, A.G. Falusi, A.P. Jarman, J.B. Clegg, and D.J. Weatherall. 1986. Analysis of the human α-globin gene cluster reveals a highly informative genetic locus. *Proc. Natl. Acad. Sci.* **83:** 5165.

Hulten, M. 1974. Chiasma distribution at diakinesis in the normal human male. *Hereditas* **76:** 55.

Jarman, A.P. and R.A. Wells. 1989. Hypervariable minisatellites: Recombinators or innocent bystanders? *Trends Genet.* **5:** 367.

Jeffreys, A.J. 1987. 23rd. Colworth medal lecture: Highly variable minisatellites and DNA fingerprints. *Biochem. Soc. Trans.* **15:** 309.

Jeffreys, A.J., R. Neumann, and V. Wilson. 1990. Repeat unit sequence variation in minisatellites: A novel source of DNA polymorphism for studying variation and mutation by single molecule analysis. *Cell* **60:** 473.

Jeffreys, A.J., M. Turner, and P. Debenham. 1991a. The efficiency of multilocus DNA fingerprint probes for individualization and establishment of family relationships, determined from extensive casework. *Am. J. Hum. Genet.* **48:** 824.

Jeffreys, A.J., V. Wilson, and S.L. Thein, 1985. Hypervariable "minisatellite" regions in human DNA. *Nature* **314:** 67.

Jeffreys, A.J., N.J. Royle, V. Wilson, and Z. Wong. 1988. Spontaneous mutation rates to new length alleles at tandem-repetitive hypervariable loci in human DNA. *Nature* **332:** 278.

Jeffreys, A.J., A. MacLeod, K. Tamaki, D.L. Neil, and D.G. Monckton. 1991b. Minisatellite repeat coding as a digital approach to DNA typing. *Nature* 354: 204.

Kasperczyk, A., N.A. Kimartino, and T.G. Krontiris. 1990. Minisatellite allele diversification: The origin of rare alleles at the HRAS1 locus. *Am. J. Hum. Genet.* 47: 854.

Kelly, R., G. Bulfield, A. Collick, M. Gibbs, and A.J. Jeffreys. 1989. Characterization of a highly unstable mouse minisatellite locus: Evidence for somatic mutation during early development. *Genomics* 5: 844.

Kwiatkowski, D.J., E.P. Henske, K. Weimer, L. Ozelius, J.F. Gusella, and J. Haines. 1992. Construction of a GT polymorphism map of human 9q. *Genomics* 12: 229.

Laurie, D.A. and M.A. Hulten. 1985. Further studies on chiasma distribution and interference in the human male. *Ann. Hum. Genet.* 49: 203.

Litt, M. and J.A. Luty. 1989. A hypervariable microsatellite revealed by in vitro amplification of a nucleotide repeat within the cardiac muscle actin gene. *Am. J. Hum. Genet.* 44: 397.

Mahtani, M.M. and H.F. Willard. 1990. Pulsed-field gel analysis of alpha satellite DNA at the human X chromosome centromere: High frequency polymorphisms and array size estimate. *Genomics* 7: 607.

Morral, N., V. Nunes, T. Casals, and X. Estevill. 1991. CA/GT microsatellite alleles within the cystic fibrosis transmembrane conductance regulator (CFTR) gene are not generated by unequal crossingover. *Genomics* 10: 692.

Nakamura, Y., C. Julier, R. Wolff, T. Holm, P. O'Connell, M. Leppert, and R. White. 1987a. Characterization of a human "midisatellite" sequence. *Nucleic Acids Res.* 15: 2537.

Nakamura, Y., M. Lathrop, P. O'Connell, M. Leppert, M.I. Kamboh, J.-M. Lalouel, and R. White. 1989. Frequent recombination is observed in the distal end of the long arm of chromosome 14. *Genomics* 4: 76.

Nakamura, Y., M. Lathrop, P. O'Connell, M. Leppert, D. Barker, E. Wright, M. Skolnick, S. Kondoleon, M. Litt, J.-M. Lalouel, and R. White. 1988. A mapped set of markers for human chromosome 17. *Genomics* 2: 302.

Nakamura, Y., M. Leppert, P. O'Connell, R. Wolff, T. Holm, M. Culver, C. Martin, E. Fujimoto, M. Hoff, E. Kumlin, and R. White. 1987b. Variable number of tandem repeat (VNTR) markers for human gene mapping. *Science* 235: 1616.

Oakey, R. and C. Tyler-Smith. 1990. Y chromosome DNA haplotyping suggests that most European and Asian men are descended from one of two males. *Genomics* 7: 325.

O'Hoy, K.L., C. Tsilfidis, M.S. Mahadevan, C.E. Neville, J. Barcelo, A.G.W. Hunter, and R.G. Korneluk. 1993. Reduction in size of the myotonic dystrophy trinucleotide repeat mutation during transmission. *Science* 259: 809.

Page, D.C., K. Bieker, L.G. Brown, S. Hinton, M. Leppert, J.-M. Lalouel, M. Lathrop, M. Nystrom-Lahti, A. De La Chappelle, and R. White. 1987. Linkage, physical mapping and DNA sequence analysis of pseudoautosomal loci on the human X and Y chromosomes. *Genomics* 1: 243.

Richards, R.I., K. Holman, K. Friend, E. Kremer, D. Hillen, A. Staples, W.T. Brown, P. Goonewardena, J. Tarleton, C. Schwartz, and G.R. Sutherland.

1992. Evidence of founder chromosomes in fragile X syndrome. *Nature Genet.* **1:** 257.

Riggins, G.J., L.K Lokey, J.L. Chastain, H.A. Leiner, S.L. Sherman, K.D. Wilkinson, and S.T. Warren. 1992. Human genes containing polymorphic trinucleotide repeats. *Nature Genet.* **2:** 186.

Rouyer, F., M.-C. Simmler, C. Johnsson, G. Vergnaud, H.J. Cooke, and J. Weissenbach. 1986. A gradient of sex linkage in the pseudoautosomal region of the human sex chromosomes. *Nature* **319:** 291.

Royle, N.J., R.E. Clarkson, Z. Wong, and A.J. Jeffreys. 1988. Clustering of hypervariable minisatellites in the proterminal regions of human autosomes. *Genomics* **3:** 352.

Royle, N.J., J.A.L. Armour, M. Webb, A. Thomas, and A.J. Jeffreys. 1992. A hypervariable locus D16S309 located at the distal end of 16p. *Nucleic Acids Res.* **20:** 1164.

Schumm, J., R. Knowlton, J. Braman, D. Barker, G. Vovis, G. Akots, V. Brown, T. Gravius, C. Helms, K. Hsiao, K. Rediker, J. Thurston, D. Botstein, and H. Donis-Keller. 1985. Detection of more than 500 single-copy RFLPs by random screening. *Cytogenet. Cell Genet.* **40:** 739.

Shelbourne, P., R. Winqvist, E. Kunert, J. Davies, J. Leisti, H. Thiele, H. Bachmann, J. Buxton, B. Williamson, and K. Johnson. 1992. Unstable DNA may be responsible for the incomplete penetrance of the myotonic dystrophy phenotype. *Hum. Mol. Genet.* **1:** 467.

Sherrington, R., G. Melmer, M. Dixon, D. Curtis, B. Mankoo, G. Kalsi, and H. Gurling. 1991. Linkage disequilibrium between two highly polymorphic microsatellites. *Am. J. Hum. Genet.* **49:** 966.

Tamaki, K., D.G. Monckton, A. MacLeod, D.L. Neil, M. Allen, and A.J. Jeffreys. 1992. Minisatellite variant repeat (MVR) mapping: analysis of "null" repeat units at D1S8. *Hum. Mol. Genet.* **1:** 401.

Tautz, D. 1989. Hypervariability of simple sequences as a general source for polymorphic markers. *Nucleic Acids Res.* **17:** 6463.

Vergnaud, G., D. Mariat, F. Apiou, A. Aurias, M. Lathrop, and V. Lauthier. 1991. The use of synthetic tandem repeats to isolate new VNTR loci: Cloning of a human hypermutable sequence. *Genomics* **11:** 135.

Weber, J.L. 1990. Informativeness of human (dA-dC)n·(dG-dT)n polymorphisms. *Genomics* **7:** 524.

Weber, J.L. and P.E. May. 1989. Abundant class of human DNA polymorphisms which can be typed using the polymerase chain reaction. *Am. J. Hum. Genet.* **44:** 388.

Weissenbach, J., G. Gyapay, C. Dib, A. Vignal, J. Morisette, P. Millasseau, G. Vaysseix, and M. Lathrop. 1992. A second-generation linkage map of the human genome. *Nature* **359:** 794.

Willard, H.F. 1991. Evolution of alpha satellite. *Curr. Opin. Genet. Dev.* **1:** 509.

Wolff, R.K., Y. Nakamura, and R. White. 1988. Molecular characterization of a spontaneously generated new allele at a VNTR locus: No exchange of flanking DNA sequence. *Genomics* **3:** 347.

Wolff, R.K., R. Plaetke, A.J. Jeffreys, and R. White. 1989. Unequal crossingover between homologous chromosomes is not the major mechanism involved in the generation of new alleles at VNTR loci. *Genomics* **5:** 382.

Wong, Z., V. Wilson, A.J. Jeffreys, and S.L. Thein. 1986. Cloning a selected fragment from a human DNA "fingerprint": Isolation of an extremely polymorphic minisatellite. *Nucleic Acids Res.* **14:** 4605.

Wong, Z., V. Wilson, I. Patel, S. Povey, and A.J. Jeffreys. 1987. Characterization of a panel of highly variable minisatellites cloned from human DNA. *Ann. Hum. Genet.* **51:** 269.

Wyman, A.R., L.B. Wolfe, and D. Botstein. 1985. Propagation of some human DNA sequences in bacteriophage λ vectors requires mutant *Escherichia coli* hosts. *Proc. Natl. Acad. Sci.* **82:** 2880.

Yu, S., J. Mulley, D. Loesch, G. Turner, A. Donnelly, A. Gedeon, D. Hillen, E. Kremer, M. Lynch, M. Pritchard, G.R. Sutherland, and R.I. Richards. 1992. Fragile-X syndrome: Unique genetics of the heritable unstable element. *Am. J. Hum. Genet.* **50:** 968.

Are Repetitive DNA Sequences Involved with Leukemia Chromosome Breakpoints?

Raymond L. Stallings,[1] Norman A. Doggett,[2] Katsuzumi Okumura[3], A. Gregory Matera,[3] and David C. Ward[3]

[1]Department of Human Genetics
University of Pittsburgh
Pittsburgh, Pennsylvania 15261

[2]Life Sciences Division
Los Alamos National Laboratory
Los Alamos, New Mexico 87544

[3]Department of Human Genetics
Yale University School of Medicine
New Haven, Connecticut 06510

Repetitive DNA sequences have been referred to as "junk" DNA (Ohno 1972). Although repetitive sequence DNA represents substantial portions of eukaryotic genomes, it has been difficult to establish a functional significance for such sequences. Rapid evolution of some repetitive sequences, such as primate α satellite, is consistent with the concept of junk DNA. Alternatively, other repetitive sequences, such as those forming functional centromeres (Grady et al. 1992) and telomeres (Moyzis et al. 1988), possess highly conserved sequences and play very important roles in cell division. It is clear, however, that both types of repetitive sequences (functional and junk) are implicated in a wide variety of human inherited diseases. It seems logical that repetitive DNA sequences capable of causing gene dysfunction, either by insertional mutagenesis or by facilitating illegitimate recombination, might also play a significant role in the neoplastic process. In this chapter, we examine the available evidence suggesting that several types of repeats, including Alu, telomere, and minisatellites, are involved in the generation of chromosome breakpoints found in leukemia.

Topics discussed include:

❏ Alu sequences and leukemia breakpoints
❏ telomeric repeats and leukemia breakpoints
❏ minisatellite repeats and leukemia breakpoints
❏ undefined repeats and leukemia breakpoints

INTRODUCTION

Several different classes of repetitive DNA sequences have been implicated in the etiology of a broad spectrum of inherited human diseases. For instance, amplification of trinucleotide microsatellite repeats is strongly associated with fragile X syndrome (Kremer et al. 1991), spinobulbar muscular atrophy (La Spada et al. 1991), myotonic dystrophy (Mahadevan et al. 1992), and Huntington's disease (Huntington's Disease Collaborative Research Group 1993). In addition, transposition of Alu and L1 interspersed repeats into genes can disrupt gene function and have led to a number of different disease phenotypes, including hemophilia (Dombroski et al. 1991), acholinesterasemia (Muratani et al. 1991), and neurofibromatosis (Wallace et al. 1991). Illegitimate homologous recombination between Alu repeats in introns of the HPRT and LDL receptor genes results in exon duplication events that cause Lesch-Nyhan syndrome (Marcus et al. 1993) and familial hypercholesterolemia (Lerhman et al. 1987).

Given the active role that repetitive DNA sequences play in the generation of some types of inherited disease, we might suspect that repeat sequences could also play a role in neoplastic diseases by facilitating chromosome rearrangements in somatic cells. Indeed, three different types of repeats, Alu, telomere, and minisatellites, have been suggested to play a role in the generation of chromosome breakpoints in leukemia.

Alu SEQUENCES AND LEUKEMIA TRANSLOCATION BREAKPOINTS

Interspersed Alu repetitive sequences may play a role in acute lymphoblastic leukemia (ALL) and chronic myeloid leukemia (CML). The most common translocation seen with these leukemias is the t(9;22)(q34;q11) that joins a section of the *ABL* proto-oncogene on chromosome 9 to the *BCR* gene on chromosome 22. This translocation is cytogenetically discernible by the presence of the 22q⁻ chromosome, the so-called Philadelphia chromosome (Ph positive; for review, see Mittleman 1986). In some instances, the translocation breakpoints appear to occur within Alu sequences found in intronic regions of the *ABL* and *BCR* genes, supporting the concept that Alu sequences might be hot spots for recombination (de Klein et al. 1986; Chen et al. 1989). There is, however, heterogeneity in the position of the breakpoint. In one Ph-positive ALL case, the breakpoint region on chromosome 9 has homology with the Alu consensus sequence, whereas the chromosome 22 breakpoint region does not (van der Feltz et al. 1989). Although recombination between Alu sequences may be responsible for some translocation events, other mechanisms are also likely to be involved.

Alu sequences are also found at the breakpoint regions in ALL and CML cases involving translocations t(4;11) and t(9;11), respectively. The chromosome 11q23 breakpoint occurs within a 5-kb region composed mostly of Alu and L1 repeats (Djabali et al. 1992). The breakpoint in one t(4;11) ALL case was further mapped to a 1.2-kb region that is composed entirely of Alu repeats (Djabali et al. 1992). It is not known whether Alu occurs at the breakpoint junctions of chromosomes 4 and 9 in these cases.

The presence of Alu sequences at translocation junctions in a number of leukemia cases suggests that these repeats play a role in the translocation process. Alu-Alu recombination has also been demonstrated to occur in an X-Y translocation, resulting in XX maleness (Rouyer et al. 1987), and it may very well be implicated in translocations observed in neoplastic cells that have not yet been characterized at the molecular level.

If Alu elements are indeed ubiquitous, with one Alu occurring every 4 kb on average (Moyzis et al. 1989), one might expect to observe Alu-Alu recombination much more frequently. It has been suggested that nonhomologous recombination is highly suppressed in normal somatic cells, so that only Alu sequences with extensive sequence similarity have a chance for recombination (Filatov et al. 1990). Many different Alu subfamilies exist that possess varying degrees of sequence similarity (see Matera et al. 1990). Thus, different Alu subfamilies may have different affinities for various proteins (Tomilin et al. 1992). Indeed, a role for Alu-binding proteins in DNA cleavage at breakpoint junctions has been proposed (van der Feltz et al. 1989). Alternatively, they may present recombinational hot spots where recombination occurs between members of the same subfamily located at different chromosomal regions.

TELOMERIC REPEATS AND LEUKEMIA BREAKPOINTS

Another repetitive sequence that may be involved with rearrangements found in leukemia is the telomeric repeat, $(TTAGGG)_n$. This is one of the most highly conserved repeats found in vertebrate genomes (Moyzis et al. 1988), and it occurs at interstitial, as well as telomeric, locations in mammalian genomes (Meyne et al. 1990; Wells et al. 1990; Ashley and Ward 1993). Telomere fusions between different human chromosomes have been observed in a number of different virally transformed cell lines and tumors (Pathak et al. 1988; Saltman et al. 1989). Although the role that these fusions play in neoplastic transformation is uncertain, these observations suggest that telomeric repeats might be hot spots for recombination. Telomeric repeats occurring at interstitial locations could be the result of ancient chromosome fusion events (Ijdo et al. 1991) that may be very similar to the types of fusions observed in neoplastic cells.

Hastie and Allshire (1989) have proposed that interstitial telomere repeats may occur at some sites of chromosome fragility. The one cloned fragile site, FRAXA, does not contain a telomeric repeat (Kremer et al. 1991), and the FRA2B is distinct from the inverted telomere repeat arrays observed interstitially in band 2q13 (Ijdo et al. 1991). Nevertheless, it is still possible that telomeric repeats could be involved with other types of fragile sites. Some fragile sites appear to be hot spots for recombination (Meulepas et al. 1991), a property associated with telomeric repeats (Ashley and Ward 1993; Ashley et al. 1993). If interstitial telomere sequences do represent some types of fragile sites, any role in the neoplastic process is still uncertain, since any link between fragile sites and nonrandom chromosome breakpoints found in human neoplasia is controversial (Sutherland 1988; Sutherland and Simmers 1988). Perhaps work on mouse chromosome 2 shows the strongest association between fragile sites and chromosome breakpoints found in leukemia. Bouffler et al. (1993) reported that interstitial telomere repeats map very close to radiation-sensitive fragile sites on mouse chromosome 2 that correspond to breakpoints found in radiation-induced acute myeloid leukemia.

MINISATELLITE REPEATS AND LEUKEMIA BREAKPOINTS

A third type of repetitive DNA sequence that may be involved with chromosome breakpoints in leukemia is the minisatellite repeat. The first minisatellite family identified by Jeffreys et al. (1985) appears to occur predominantly in subterminal regions of human chromosomes (Royle et al. 1988). Subsequently, numerous other families of minisatellite repeats have been identified (see, e.g., Vassart et al. 1987). Some, such as the M13 family of minisatellites, appear to have sequence motifs that occur on all human chromosomes (Christmann et al. 1991), whereas others may be chromosome-specific and highly clustered within a single band of a chromosome (Buroker et al. 1987; Das et al. 1987). The minisatellite observed near the apolipoprotein gene occurs at approximately 60 sites on chromosome 19, all within band q13.3 (Das et al. 1987).

In general, minisatellites possess a core repeat between 14 bp and 40 bp in length. Perhaps the most notable feature of minisatellite loci is the hypervariability associated with the number of copies of core repeat units found at the same locus in different individuals (Jeffreys 1987). Detailed analysis of markers flanking minisatellite loci has indicated that unequal crossing-over between homologous chromosomes is not the major mechanism involved in the generation of hypervariability at these loci (Wolff et al. 1989). Although the mechanism responsible for hypervariability is not fully understood, replication slippage or unequal sister chromatid exchange during meiosis or mitosis might be involved with the hypervariability found at some minisatellite loci. In certain experimental systems, some minisatellite sequences have been observed to possess highly recombinogenic properties (Wahls et al. 1990).

Krowcyznska et al. (1990) identified a minisatellite sequence that occurs near the breakpoint regions of the MYC and BCL2 oncogene translocations. The consensus core repeat is very similar to *chi*, the prokaryotic activator of recombination. The authors suggest that this *chi*-like sequence mediates site-specific recombination between these oncogenes and immunoglobin loci in lymphocyte-derived neoplasms; however, no direct experimental evidence is available to support this hypothesis.

UNDEFINED REPEATS AND LEUKEMIA BREAKPOINTS

A number of low-abundance repeats have been identified in the human and mouse genomes that have very large repeat units. For example, mouse chromosome 1 contains approximately 50 copies of a 50-kb-long repeat unit (Purmann et al. 1992), and mouse chromosome 8 contains 60–80 copies of a repeat known to be more than 15 kb in length (Boyle and Ward 1992). We propose to call these types of repeats "macrosatellites," which differ from minisatellite repeats in the length of the repeat unit. Some macrosatellite repeats may be amplified gene sequences.

At least one macrosatellite sequence, with an 18-kb repeat unit, appears to occur on both sides of a chromosome-17 breakpoint region found in acute promyelocytic leukemia (APL) (Moore et al. 1989). These sequences appear to be clustered in a 2-Mb region at 17q11.2-12. Whether or not these sequences are involved with APL remains to be determined. Additional undefined repeats, mapping to 17q21 and 17q23, could also be involved with a complex Philadelphia chromosome translocation t(8;22;10;17)(q34;q11;p13;q21) found in one CML case (McKeithan et al. 1992).

A complex family of chromosome-16-specific low-abundance repetitive DNA sequences have also been discovered on human chromosomes 16p13, p12, p11, and q22 (ANLL) (Dauwerse et al. 1992; Stallings et al. 1992a). Their locations at 16p13 and 16q22 immediately suggested a possible involvement with acute nonlymphocytic leukemia (ANLL). In the ANLL M4 subtype, several abnormalities of chromosome 16 are prominent. A pericentric inversion with breakpoints in bands p13 and q22 is the most common abnormality (LeBeau et al. 1983). Other rearrangements affecting chromosome 16 in ANLL cases include deletion of band q22 (Arthur and Bloomfield 1983), as well as inter-homolog translocation between bands p13 and q22 (Hogge et al. 1984). The breakpoints for the inversion 16 and translocation (16p13;q22) appear to occur within the same subregion of the chromosome (Wessels et al. 1991a). In addition, translocation between band 16p13 and chromosome 8p11 has been reported in both M4 and M5 ANLL subtypes (Lai et al. 1987; Powell et al. 1988;). However, the breakpoints on 16p13 in M4 and M5 subtypes are at distinctly different locations (Wessels et al. 1991b).

Characterization of chromosome-16-specific low-abundance repeats

Chromosome-16-specific low-abundance repeats (CH16LARs) were initially discovered during the course of developing a cosmid contig physical map of chromosome 16 (Stallings et al. 1990, 1992b) because they greatly complicated the ordering of cosmid clones into contigs for some regions of chromosome 16 (Stallings et al. 1992a). The cosmids in this map represent approximately 84% of the chromosome (Stallings et al. 1992b). CH16LARs are interspersed over megabase-size regions of chromosome 16, as determined by both contig mapping and pulsed field gel electrophoresis analysis (Fig. 1) (Stallings et al. 1992). One unusually large contig, contig 55, was particularly troublesome. Contig 55 contained 74 clones, or approximately 2% of all the cosmid clones contained within the contig physical map of chromosome 16. Multiple macrorestriction fragments, totaling up to 6.6 Mbp, are observed on blots when CH16LAR-containing cosmids are used as probes. However, it is difficult to accurately define the size of the regions that contain these repeats because of effects of methylation on restriction enzyme digestion and because much of the hybridizing DNA is located in the compression zone of the pulsed field gels (Fig. 1).

CH16LAR sequences are also found in a considerable number of YAC clones derived from a 1X chromosome-16-specific YAC library (McCormick et al. 1993). Approximately 6% (29 out of 450) of the YAC clones were strongly positive following hybridization of CH16LAR-containing cosmid inserts to gridded arrays of YACs. Since the average insert size of the YAC library is 205 kb, we would estimate that these

Figure 1 Southern blot of a pulsed field gel containing BssHII, NotI, MluI, NruI, SaII, SfiI, NaeI, and NarI digested DNA (1–7, respectively) from CY18, a monochromosome 16 somatic cell hybrid. Size standards are λ concatamers in the first and last lanes (M). Cosmid clone 41E11 from contig 55 was labeled with ^{32}P, preannealed to human Cot1 fractionated DNA to mask middle repetitive DNA sequences (predominantly Alu and L1), and hybridized to the DNA on the filter membrane. Preannealing with Cot1 DNA does not mask low-abundance CH16LAR repeats, as multiple macro-restriction fragments are observed. In several digests, much of the hybridizing material is in the compression zone of the gel.

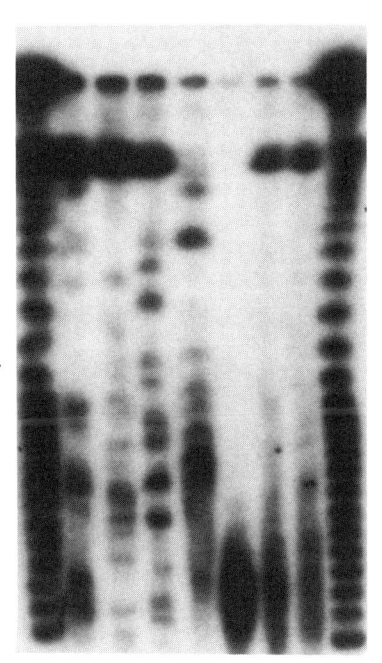

CH16LAR-containing YACs cover approximately 6 Mb of chromosome 16. It is interesting to note that, unlike some types of repetitive sequences (Neil et al. 1990), CH16LAR sequences seem stable in YAC vectors.

Inter-Alu polymerase chain reaction (PCR) products from these CH16LAR-positive YAC clones, when hybridized to the arrayed cosmid contigs, hybridize extensively to contig 55 clones. A large number of contig 55 clones (40–60) are always positive when hybridized to inter-Alu PCR products from CH16LAR-bearing YACs. Figure 2 shows a representative grid hybridization. In addition to cosmid clones from contig 55 being positive, cosmids from a number of other contigs are also positive, thus suggesting that CH16LAR sequences extend beyond the regions covered by contig 55 clones.

Clones from contig 55 have an unusually large number of small restriction fragments (1-kb to 3-kb fragments are ~3-fold greater than what are found in the average chromosome-16 cosmid clone; see Fig. 3). Some types of CH16LAR sequences appear to be highly localized within some cosmids (Fig. 3). Contig 55 clones could not be assembled into a linear contig because of the large number of clones with false overlap. If the overabundant small restriction fragments are deleted from contig 55 clone fingerprints and the clone analysis is based on more unique restriction fragments, contig 55 could be broken into several smaller contigs and singletons (single cosmids).

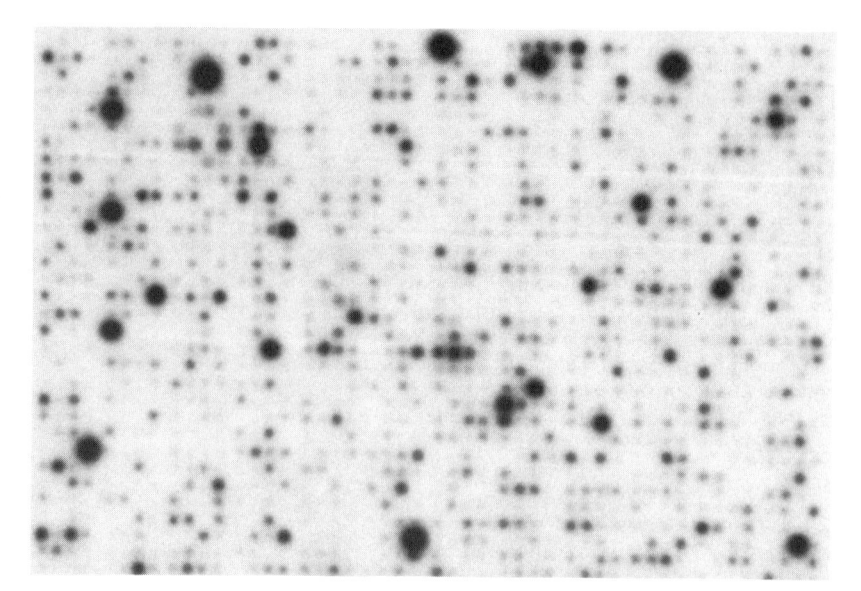

Figure 2 Typical grid hybridization of inter-Alu PCR products from a YAC containing CH16LAR sequences (clone Y16.7) to cosmid clones. 1536 cosmids (out of a total of 4000 fingerprinted clones) are gridded onto this microtiter dish size membrane. 20 cosmids, predominantly from contig 55, are strongly positive. Inter-Alu PCR products hybridize to a single restriction fragment when hybridized to Southern blots containing contig 55 clones (see Fig. 4).

Confirmation that contig 55 clones contain low-abundance repeats was provided by fluorescence in situ hybridization (FISH). A total of five clones from contig 55 were labeled with biotin dUTP, preannealed to *Cot*1-fractionated human DNA to mask repetitive DNA sequences, and hybridized to normal human metaphase chromosomes. The FISH signals were unusually intense, and some clones mapped to as many as three locations on chromosome 16 (Fig. 4). In situ hybridization results also suggest that there are several distinct types of CH16LAR sequences. For example, four distinct regions of chromosome 16 show FISH signal with contig 55 cosmids, yet any single cosmid tested hybridizes to a maximum of only three regions. In addition, cosmids containing CH16LAR sequences, when preannealed to other CH16LAR-containing cosmids, fail to suppress multiple FISH signals (Stallings et al. 1992a).

In addition to mapping contig 55 clones by FISH, we have also regionally mapped these clones using a hybrid mapping panel that contains 51 rearranged chromosomes 16 (Callen et al. 1992; Stallings et al. 1993). The average distance between breakpoints is 1.6 Mb (Callen et al. 1992), providing greater resolution than previous FISH results. Figure 5 summarizes the mapping of CH16LAR sequences with respect to the breakpoints found in this mapping panel. Regional mapping using the somatic cell hybrids was very consistent with results obtained by FISH.

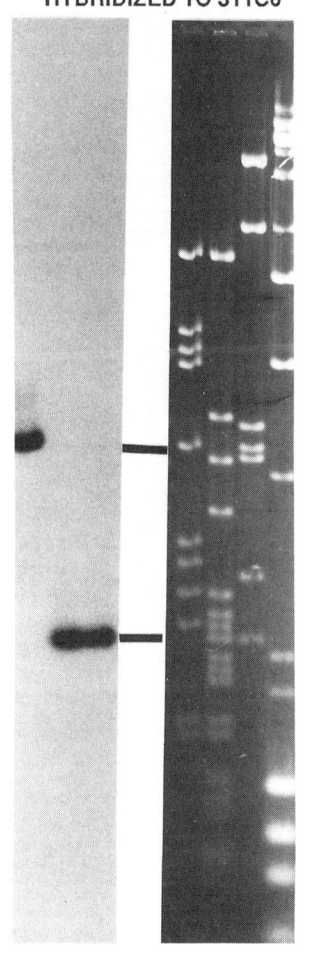

Figure 3 Hybridization of inter-Alu PCR products from Y16.7 to a Southern blot containing EcoRI (*left* lane), EcoRI/HindIII (*middle* lane), and HindIII (*right* lane) digested DNA from cosmid clone 311C6 (from contig 55). Right panel is the ethidium bromide stained gel. A single fragment, containing a CH16LAR sequence, showed positive hybridization signal on a single restriction fragment in all three digests. The HindIII-positive fragment was cloned and sequenced. (Reprinted, with permission, from Stallings et al. 1992a.)

Characterization of a 2.2-kb subcloned DNA fragment containing CH16LAR sequences revealed the presence of a minisatellite type sequence which did not possess homology with any of the known minisatellites (Fig. 6) (Stallings et al. 1992a). The plasmid containing the 2.2-kb sequence, p311C6-H2.2, hybridizes to multiple locations on chromosome 16 when used as a FISH probe (Fig. 7A). A 40-bp consensus repeat sequence identified in p311C6-H2.2 is also present in a large number of contig 55 clones, and quantitative slot blot analysis indicates that 250 copies exist in the human genome. Hybridization of a 31-nucleotide consensus oligomer results in multiple FISH signals in the 16p12-13 region (Fig. 7B), but no signal was detected on the q arm. The most likely explanation is that the q arm contains too few of these

CONTIG 55 CLONES POSITIONED

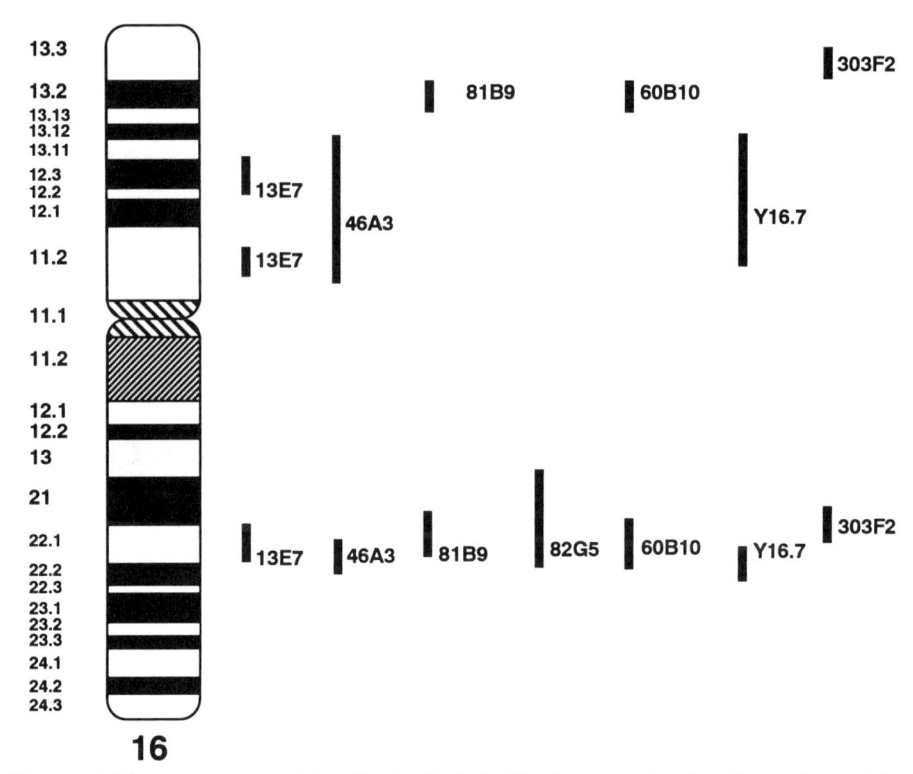

Figure 4 Ideogram summarizing the in situ hybridization mapping locations of cosmid clones from contig 55 and YAC clone Y16.7. (Reprinted, with permission, from Stallings et al. 1992a.)

repeats to be detected by FISH. Alternatively, it is also possible that a small, unsequenced region from p311C6-H2.2 contains additional repeats that map to the p and q arms; approximately 10–15% of p311C6-H2.2 was not sequenced because these regions appeared to be composed entirely of Alu sequences (Stallings et al. 1992a).

Whether the majority of CH16LAR sequences are also short minisatellite-type sequences, or "macrosatellite" sequences, is unknown. It is interesting to note that mouse chromosome 8, which possesses ex-

Figure 5 Ideograms of chromosome 16p (A) and 16q (B) showing the relative order of chromosome 16 breakpoints in different somatic cell hybrids. The regions where five different contig-55 clones map are indicated with vertical arrows. Coding sequences that have been mapped by this hybrid mapping panel are also indicated, along with the ANLL inversion breakpoints. (Reprinted, with permission, from Stallings et al. 1993.)

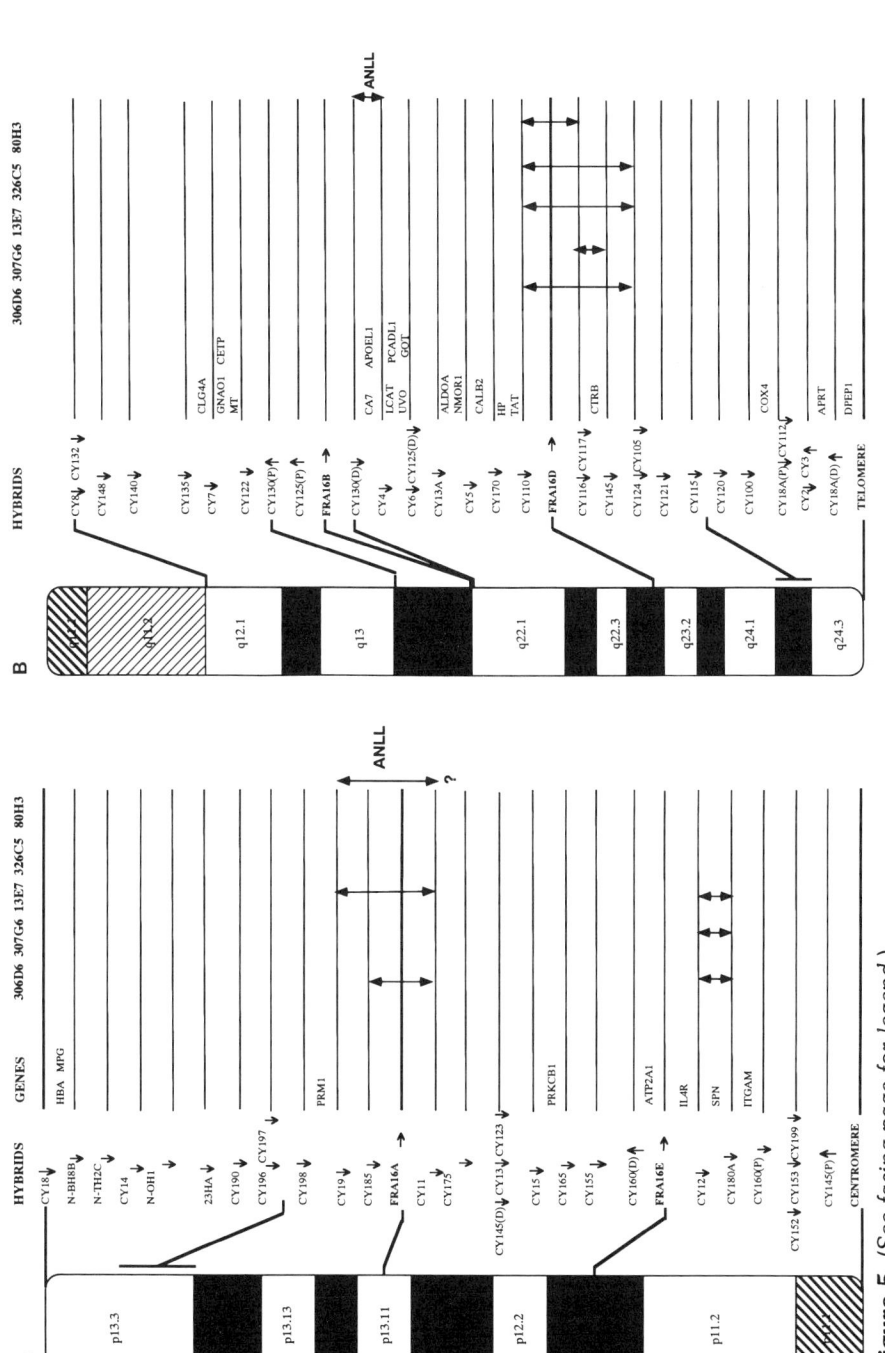

REPETITIVE SEQUENCES AND LEUKEMIA BREAKPOINTS 69

Figure 5 (See facing page for legend.)

```
1)              TCCT  TCA    TCCC   CTTCCACCCT  CAGTGGATGA  TAATATCAAG GA
2)        GTG   TCCT  CTTGC  TCCT   CTTCCACCCT  CAGTGGATGA  TAATCTGAAG GA
3)        GTATC TCCT  GG     TCCCT  CTTCCACCCT  C
4)                           TCCT   CTTCCACCCT  CAGTGGATGA  TAATCTGAAG GA
5)  CTGTCTCTTTG TCCCT CTTCCACCC TCCT CTTCCACCCT  CAG-TGGATG  ATAATCTCAA -GA
```
                                                          G
Consensus Repeat = TCCT X TCCT CTTCCACCCT CAGTGGATGA TAATCT AAG GA
                                                          C

Figure 6 The minisatellite repeat locus found in p311C6-H2.2. The exact order of nucleotides is shown on lines 1–5. Each line is aligned to show the repeats. The consensus repeat sequence is also shown.

tensive homology with human chromosome 16 (Scherer et al. 1989), also contains a very large macrosatellite repeat. The repeats found on mouse chromosome 8 do not hybridize to human DNA. The fact that CH16LAR sequences greatly complicated contig analysis using a restriction enzyme-based fingerprinting method (Stallings et al. 1990) suggests that some of these repeats could be quite large.

The p311C6-H2.2 contains at least five distinct Alu repeats, and the minisatellite sequence found in p311C6-H2.2 occurs in a highly Alu-rich region. This is very consistent with the observation that CH16LAR sequences are amplified from a large number of YACs by inter-Alu PCR. Other types of minisatellite repeats (Armour et al. 1989), as well as some types of microsatellite repeats (Beckman and Weber 1992), also appear to be clustered in regions that are dense in other types of interspersed repeats.

Are CH16LAR sequences involved with ANLL breakpoints?

It is not unreasonable to consider that CH16LAR sequences may be causally related to the inversions and translocations that occur in leukemia cells, since two of the regions where contig 55 clones map, p13 and q22, appear to coincide with the positions of pericentric inversion, deletion, and translocation breakpoints found in ANLL cells. Dauwerse et al. (1992) have independently identified cosmid clones that produce similar results to contig 55 clones following in situ hybridization. In addition, they observed that one clone (cosmid 177) hybridizes very intensely to two blocks on the p arm and very weakly to one block on the q arm. Cosmid 177 also gives a very weak signal on chromosome 18p. Interestingly, the cosmid 177 signal on the distal p arm block is transferred to the q arm of the inversion chromosome 16 found in ANLL cells. This indicates that the low-abundance repeats in cosmid 177 flank the inversion breakpoint on the p arm (Dauwerse et al. 1992).

The somatic cell hybrid mapping of contig 55 clones indicates that CH16LAR sequences in at least two cosmid clones map in close proxim-

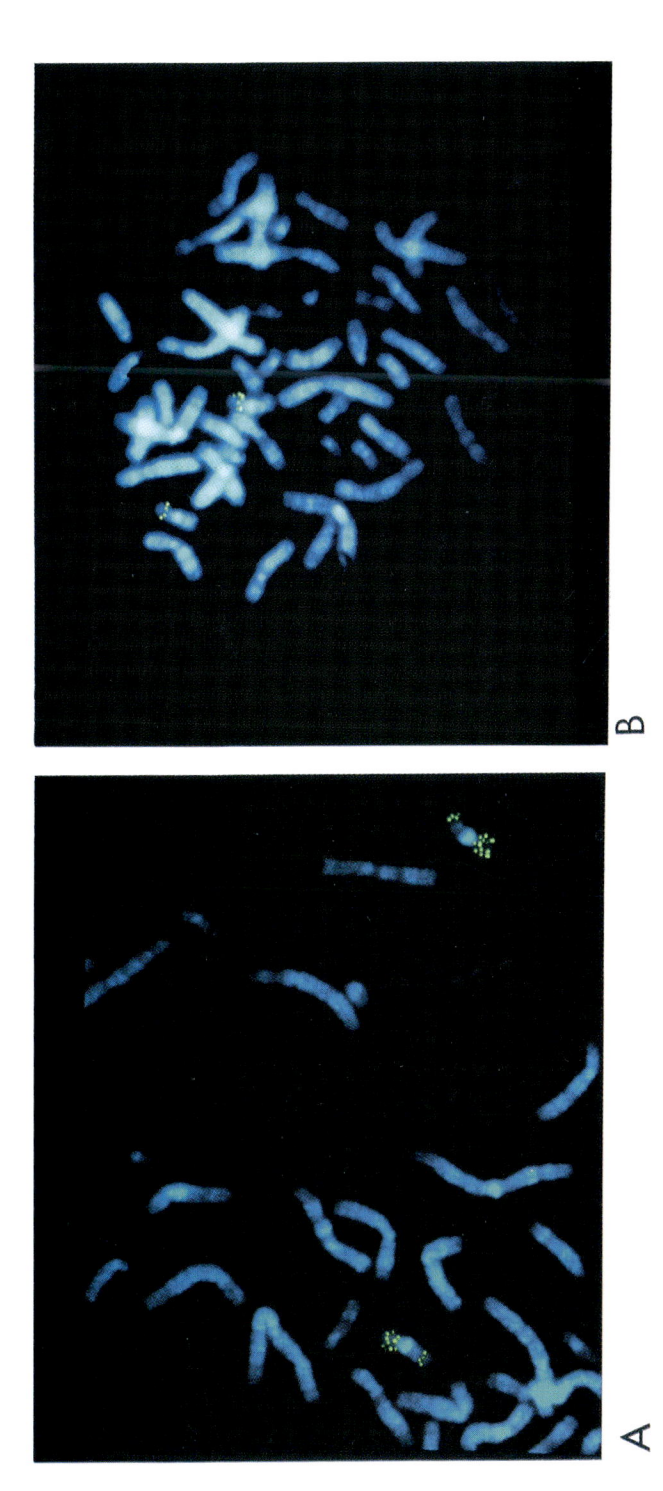

Figure 7 FISH using a biotin-labeled p311C6-H2.2 as probe (A) and (B) a biotinylated 31-nt oligomer corresponding to the core of the consensus sequence shown in Fig. 6. Chromosomes were counterstained with DAPI.

ity to the ANLL p-arm breakpoint. However, they show more intense FISH signal on 16q than cosmid 177 and are thus more like the repeats found in cosmid 163 reported by Dauwerse et al. (1992). In situ hybridization of one of these clones, 13E7, to metaphases from ANLL cells indicates that CH16LAR sequences found in this cosmid clone map proximal to the p arm ANLL breakpoint (R.L. Stallings, unpubl.). This fact, coupled with the fact that 13E7 does not hybridize to 18p, indicates that the repeats in 13E7 are distinctly different from those found in cosmid 177. However, at least one clone from contig 55, 36G7, appears to contain repeats similar to those found in cosmid 177. FISH analysis with cosmid 36G7 on ANLL subtype M4 metaphase chromosomes shows the transfer of p-arm CH16LAR sequences to the q arm on the inversion chromosome 16 (R.L. Stallings, unpubl.).

On 16q, the positions of all of the repeats mapped by in situ hybridization by Stallings et al. (1992a) and Dauwerse et al. (1992) appear to be identical. By somatic cell hybrid breakpoint mapping, however, the ANLL inversion breakpoint maps between the loci for D16S4 and D16S91 (between the breakpoints found in hybrids CY130(D) and CY4 [Baker et al. 1990; Callen et al. 1992]). This region is definitely proximal to where the highest density of CH16LAR sequences map (see Fig. 5); however, the exact distance between the ANLL breakpoint and the CH16LAR dense region is unknown. Although this distance is too small to be discernible by metaphase FISH, nine genes have been mapped to the interval between the ANLL q-arm breakpoint and the region where the high density of CH16LAR sequences map (Fig. 5B). It is possible that the repeats in cosmid 177 map closer to the q-arm breakpoint region than any of the contig 55 clones that have been placed on the hybrid breakpoint map. The fact that an unknown number of different repeat elements comprise the CH16LARs makes it difficult to determine which repeat, if any, is involved with the breakpoint. However, it seems quite likely that one of the repeats resides in the vicinity of the inversion and translocation breakpoint.

As noted on the ideogram in Figure 5, CH16LAR repeats appear to map in close proximity to three out of four fragile sites found on chromosome 16, so that CH16LAR sequences may be associated with other types of chromosome 16 instability. The fragile site nearest the ANLL breakpoint (FRA16B), however, does not appear to have CH16LAR sequences in close proximity.

Pericentric inversion of chromosome 16 involving bands 16p13 and q22 has also been noted in the germ line of at least three individuals (Ionasescu et al. 1987; Bianchi et al. 1992), and we are left with the question of whether or not CH16LAR sequences have mediated these germ-line inversions. Precise molecular mapping of these germ-line inversion breakpoints with respect to the breakpoints found in ANLL will be of interest.

CONCLUDING REMARKS

Chromosome architecture appears to be riddled with a vast array of low-abundance, chromosome-specific repetitive DNA sequences. Chromosome 16 is heavily laden with a variety of repeats that include a large block of chromosome-16-specific, variant satellite II DNA in band q11.2 (Moyzis et al. 1987); a chromosome-16-specific α-satellite variant sequence at the centromere (Greig et al. 1989); and a large block of CH16LAR sequences in bands 16p13, p12, and p11.2 and q22 (Dauwerse et al. 1992; Stallings et al. 1992a). The region 16p11.2 appears to be particularly rich in low-abundance repeats that map to a limited number of other human chromosomes. A polymorphic minisatellite repeat (MS29) maps to this region as well as to chromosome 6 (Wong et al. 1990), and we have identified sets of cosmid clones that map to 16p11.2 and to either chromosome 14qter or chromosome 15q11.2 (K. Okumura et al., in prep.). It is possible that repetitive sequences found at 16p11.2 are responsible for a number of chromosome 16 rearrangements, including apparent amplification of the region 16p11.2 (Bryke et al. 1990) and duplication of 16p11-p13 (Cohen et al. 1983).

The association of repetitive DNA sequences with leukemia chromosome breakpoints is not as strong as the association between some types of repeats and several inherited diseases. However, sequences capable of transposition, amplification, or enhanced recombination are likely to play a major role in many different types of malignancies by altering the expression of oncogenes and/or tumor suppressor genes.

References

Armour, J.A., Z. Wong, V. Wilson, N.J. Royle, and A.J. Jeffreys. 1989. Sequences flanking the repeat arrays of human minisatellites: Association with tandem and dispersed repeat elements. *Nucleic Acids Res.* **17:** 4925.

Arthur, D.C. and C.D. Bloomfield. 1983. Partial deletion of the long arm of chromosome 16 and bone marrow eosinophilia in acute nonlymphocytic leukemia, a new association. *Blood* **61:** 994.

Ashley, T. and D.C. Ward. 1993. A hot spot of recombination coincides with interstitial telomeric sequence in the Armenian hamster. *Cytogenet. Cell Genet.* **62:** 169.

Ashley, T., N.L.A. Cacheiro, L.B. Russell, and D.C. Ward. 1993. Molecular characterization of a pericentric inversion in mouse chromosome 8 implicates telomeres as promoters of meiotic recombination. *Chromosoma* **102:** 112.

Baker, E., D.F. Callen, D.M. Garson, and A.K. West. 1990. The human metallothionein gene cluster is not disrupted in myelomonocytic leukemia. *Genomics* **6:** 144.

Beckman, J.S. and J.L. Weber. 1992. Survey of human and rat microsatellites. *Genomics* **12:** 627.

Bianchi, D.W., R.D. Nichols, K.A. Russell, W.A. Miller, M. Ellin, and J.M. Lage. 1992. Pericentric inversion of chromosome 16 in a large kindred: Spectrum of morbidity and mortality in offspring. *Am. J. Med. Genet.* **43**: 791.

Bouffler, S., A. Silver, D. Papworth, J. Coates, and R. Cox. 1993. Murine radiation myeloid leukaemogenesis: Relationship between interstitial telomere-like sequences and chromosome 2 fragile sites. *Genes Chromosomes Cancer* **6**: 98.

Boyle, A.L. and D.C. Ward. 1992. Isolation and initial characterization of a large repeat sequence element specific to mouse chromosome 8. *Genomics* **12**: 517.

Bryke, C.R., W.R. Breg, R.P. Venkateswara, and T.L. Yang-Feng. 1990. Duplication of euchromatin without phenotypic effects: A variant of chromosome 16. *Am. J. Med. Genet.* **36**: 43.

Buroker, H., R. Bestwick, G. Haight, R.E. Magenis, and M. Litt. 1987. A hypervariable repeated sequence on human chromosome 1p36. *Hum. Genet.* **77**: 175.

Callen, D.F., N.A. Doggett, R.L. Stallings, S.A. Whitmore, S.A Lane, J. Nancarrow, L.Z. Chen, S. Apostolou, A.D. Thompson, E. Baker, S. Yang, and G.R. Sutherland. 1992. A high resolution cytogenetic-based physical map of human chromosome 16. *Genomics* **13**: 1178.

Chen, S.J., Z. Chen, M.-P. Font, L. d'Auriol, C.-J. Larsen, and R. Berger. 1989. Structural alterations of the BCR and ABL genes in Ph1 positive acute leukemias with rearrangements in the BCR gene first intron: Further evidence implicating Alu sequences in the chromosome translocation. *Nucleic Acids Res.* **17**: 7631.

Christmann, A., P.J. Lagoda, and K.D. Zang. 1991. Nonradioactive *in situ* hybridization pattern of the M13 minisatellite sequences on human metaphase chromosomes. *Hum. Genet.* **86**: 487.

Cohen, M.M., C.Lerner, and N.E. Balkin. 1983. Duplication of 16p from insertion of 16p into 16q with subsequent duplication due to crossing over within the inserted segment. *Am. J. Med. Genet.* **14**: 89.

Das, H.K., C.L. Jackson, D.A. Miller, T. Leff, and J.L. Breslow. 1987. The human apolipoprotein C-II gene sequence contains a novel chromosome 19-specific minisatellite in its third intron. *J. Biol. Chem.* **262**: 4787.

Dauwerse, J.G., E.A. Jumelet, J.W. Wessels, J.J. Saris, A. Hagemeijer, G.C. Beverstock, G.J.B. van Ommen, and M.H. Breuning. 1992. Extensive cross-homology between the long and short arm of chromosome 16 may explain leukemic inversions and translocations. *Blood* **79**: 1299.

de Klein, A., T. van Agthoven, C. Groffen, N. Heisterkamp, J. Groffen, and G. Grosveld. 1986. Molecular analysis of both translocation products of a Philadelphia-positive CML patient. *Nucleic Acids Res.* **14**: 7071.

Djabali, M., L. Selleri, P. Parry, M. Bower, B.D. Young, and G.A. Evans. 1992. A trithorax-like gene is interrupted by chromosome 11q23 translocations in acute leukemias. *Nature Genet.* **2**: 113.

Dombroski, B.A., S.L. Mathias, E. Nanthakumar, A.F. Scott, and H.H. Kazazian. 1991. Isolation of an active human transposable element. *Science* **254**: 1805.

Filatov, L.V., S.E. Mamayeva, and N.V. Tomilin. 1991. Alu family variations in neoplasia. *Cancer Genet. Cytogenet.* **56**: 11.

Grady, D.L., R.L. Ratliff, D.L. Robinson, E.C. McCanlies, J. Meyne, and R.K.

Moyzis. 1992. Highly conserved repetitive DNA sequences are present at human centromeres. *Proc. Natl. Acad. Sci.* **89**: 1695.

Greig, G.M., S.B. England, H.M. Bedford, and H.F. Willard. 1989. Chromosome-specific alpha satellite DNA from the centromere of human chromosome 16. *Am. J. Hum. Genet.* **45**: 862.

Hastie, N.D. and R.C. Allshire. 1989. Human telomeres: Fusion and interstitial sites. *Trends Genet.* **5**: 326.

Hogge, D.E., S. Misawa, N.Z. Parsa, A. Pollak, and J.R. Testa. 1984. Abnormalities of chromosome 16 in association with acute myelomonocytic leukemia and dysplastic bone marrow eosinophils. *J. Clin. Oncol.* **2**: 550.

Huntington's Disease Collaborative Research Group. 1993. A novel gene containing a trinucleotide repeat that is expanded and unstable on Huntington's disease chromosomes. *Cell* **72**: 971.

Ijdo, J.W., A. Baldini, D.C. Ward, S.T. Reeders, and R.A. Wells. 1991. Origin of human chromosome 2: An ancestral telomere-telomere fusion. *Proc. Natl. Acad. Sci.* **88**: 9051.

Ionasescu, V., S. Patil, M. Hart, W. Rhead, and I.F. Smith. 1987. Multiple congenital anomalies syndrome with myopathy in chromosome 16 abnormality. *Am. J. Med. Genet.* **26**: 189.

Jeffreys, A.J. 1987. Highly variable minisatellites and DNA fingerprints. *Biochem. Soc. Trans.* **15**: 309.

Jeffreys, A.J., V. Wilson, and S.L. Thein. 1985. Hypervariable "minisatellite" regions in human DNA. *Nature* **314**: 67.

Kremer, E.J., M. Pritchard, M. Lynch, S. Yu, K. Holman, E. Baker, S.T. Warren, D. Schlessinger, G.R. Sutherland, and R.I. Richards. 1991. Mapping of DNA instability at the fragile X to a trinucleotide repeat sequence p(CCG)n. *Science* **252**: 1711.

Krowczynska, A.M., R.A. Rudders, and T.G. Krontiris. 1990. The human minisatellite consensus at breakpoints of oncogene translocations. *Nucleic Acids Res.* **18**: 1121.

Lai, J.L., M. Zandecki, J.P. Jouet, J.B. Savary, A. Lambiliotte, F. Bouters, A. Cosson, and M. Demiatti. 1987. Three cases of translocation (8;16)(p11;p13) observed in acute myelomonocytic leukemia. *Cancer Genet. Cytogenet.* **27**: 101.

La Spada, A.R., E.M. Wilson, D.B. Lubahn, A.E. Harding, and K.H. Fischbeck. 1991. Androgen receptor gene mutations in X-linked spinal and bulbar muscular atrophy. *Nature* **352**: 77.

LeBeau, M.M., R.A. Larson, M.A. Bitter, J.W. Vardiman, H.M. Golomb, and J.D. Rowley. 1983. Association of an inversion of chromosome 16 with abnormal marrow eosinophils in acute myelomonocytic leukemia: A unique cytogenetic-clinicopathologic association. *N. Engl. J. Med.* **309**: 630.

Lerhman, M.A., J.L. Goldstein, D.W. Russell, and M.S. Brown. 1987. Duplication of seven exons in LDL receptor gene caused by Alu-Alu recombination in a subject with familial hypercholesterolemia. *Cell* **48**: 827.

Mahadevan, M., C. Tsilfidis, L. Sabourin, G. Shutler, C. Amemiya, G. Jansen, C. Neville, M. Narang, J. Barcelo, K. O'Hoy, S. Leblond, J. Earle-MacDonald, P.J. DeJong, B. Wieringa, and R.G. Korneluk. 1992. Myotonic dystrophy mutation: An unstable CTG repeat in the 3' untranslated region of the gene. *Science* **255**: 1253.

Marcus, S., D. Hellgren, B. Lambert, S.P. Fallstrom, and J. Wahlstrom. 1993.

Duplication in the hypoxanthine phosphoribosyl-transferase gene caused by Alu-Alu recombination in a patient with Lesch Nyhan syndrome. *Hum. Genet.* **90**: 477.

Matera, A.G., U. Hellman, and C.W. Schmid. 1990. A transpositionally and transcriptionally competent Alu subfamily. *Mol. Cell. Biol.* **10**: 5424.

McCormick, M.K., E. Campbell, L.L. Deaven, and R. Moyzis. 1993. Low frequency chimeric yeast artificial chromosome libraries from flow sorted human chromosomes 16 and 21. *Proc. Natl. Acad. Sci.* **90**: 1063.

McKeithan, T.W., L. Warshawsky, R. Espinosa, and M.M. LeBeau. 1992. Molecular cloning of the breakpoints of a complex Philadelphia chromosome translocation: Identification of a repeated region on chromosome 17. *Proc. Natl. Acad. Sci.* **89**: 4923.

Meulepas, L.E., J.P. Fryns, H. Van den Berghe, and J.J. Cassiman. 1991. Spontaneous FRA16B is a hotspot for sister chromatid exchanges. *Hum. Genet.* **87**: 583.

Meyne, J., R.J. Baker, H.H. Hobart, T.C. Hsu, O.A. Ryder, O.G. Ward, J.E. Wiley, D.H. Wurster-Hill, T.L. Yates, and R.K. Moyzis. 1990. Distribution of nontelomeric sites of the (TTAGGG)n telomeric sequence in vertebrate chromosomes. *Chromosoma* **99**: 3.

Mittleman, F. 1986. Clustering of breakpoints to specific chromosomal regions in human neoplasia. A survey of 5,345 cases. *Hereditas* **104**: 113.

Moore, G., P.J. Hedge, S.H. Rider, W. Xu, S. Hing, R. Palmer, D. Sheer, and E. Solomon. 1989. Multiple tandem 18-kb sequences clustered in the region of the acute promyelocytic leukemia breakpoint on chromosome 17. *Genomics* **4**: 152.

Moyzis, R.K., D.C. Torney, J. Meyne, J.M. Buckingham, J.R. Wu, C. Burks, K.M. Sirotkin, and W.B. Goad. 1989. The distribution of interspersed repetitive DNA sequences in the human genome. *Genomics* **4**: 273.

Moyzis, R.K., J.M. Buckingham, L.S. Cram, M. Dani, L.L. Deaven, M.D. Jones, J. Meyne, R.L. Ratliff, and J.R. Wu. 1988. A highly conserved repetitive DNA sequence, (TTAGGG)n, present at the telomeres of human chromosomes. *Proc. Natl. Acad. Sci.* **85**: 6622.

Moyzis, R.K., K.L. Albright, M.F. Bartholdi, L.S. Cram, L.L. Deaven, C.E. Hildebrand, N.E. Joste, J.L. Longmire, J. Meyne, and T. Schwarzacher-Robinson. 1987. Human chromosome-specific repetitive DNA sequences: Novel markers for genetic analysis. *Chromosoma* **95**: 375.

Muratani, K., T. Hada, Y. Yamamoto, T. Kaneko, Y. Shigeto, T. Ohue, J. Furuyama, and K. Higashino. 1991. Inactivation of the cholinesterase gene by Alu insertion: Possible mechanism for human gene transposition. *Proc. Natl. Acad. Sci.* **88**: 11315.

Neil, D.L., A. Villasante, R.B. Fisher, D. Vetrie, B. Cox, and C. Tyler-Smith. 1990. Structural instability of human tandemly repeated DNA sequences cloned in yeast artificial chromosomes vectors. *Nucleic Acids Res.* **18**: 1421.

Ohno, S. 1972. So much "junk" DNA in our genomes. In *Evolution of genetic systems* (ed. H.H. Smith), p. 366. Gordon and Breach, New York.

Pathak, S., Z. Wang, M.K. Dhaliwal, and P.C. Sacks. 1988. Telomeric association: Another characteristic of cancer chromosomes. *Cytogenet. Cell Genet.* **47**: 227.

Powell, B.L., J.W. McNay, S. Brown, M.R. Cooper, and M.J. Pettenali. 1988. Translocation (8;16)(p11;p13) in patients with acute monocytic leukemia.

An evolving syndrome. *Cancer Genet. Cytogenet.* **36:** 109.
Purmann, L., C. Plass, M. Gruneberg, H. Winking, and W. Traut. 1992. A long range repeat cluster in chromosome 1 of the house mouse, *Mus musculus*, and its relation to a germline homogeneously staining region. *Genomics* **12:** 80.
Rouyer, F., M.-C. Simmler, D.C. Page, and J. Weissenbach. 1987. A sex chromosome rearrangement in a human XX male caused by Alu-Alu recombination. *Cell* **51:** 417.
Royle, N.J., R.E. Clarkson, Z. Wong, and A.J. Jeffreys. 1988. Clustering of hypervariable minisatellites in the proterminal regions of human autosomes. *Genomics* **3:** 352.
Saltman, D., F.M. Ross, J.A. Fantes, R. Allshire, G.E. Turner, and H.J. Evans. 1989. Telomeric associations in a lymphoblastoid cell line from a patient with B-cell follicular lymphoma. *Cytogenet. Cell Genet.* **50:** 230.
Scherer, G., E. Bausch, A. Gaa, and O. von Deimling. 1989. Gene mapping on mouse chromosome 8 by interspecific crosses: New data on a linkage group conserved on human chromosome 16q. *Genomics* **5:** 275.
Stallings, R.L., N.A. Doggett, K. Okumura, and D.C. Ward. 1992a. Chromosome 16 specific repetitive DNA sequences that map to chromosomal regions known to undergo breakage/rearrangement in leukemia cells. *Genomics* **13:** 332.
Stallings, R.L., S.A. Whitmore, N.A. Doggett, and D.F. Callen. 1993. Refined physical mapping of chromosome 16-specific low abundance repetitive DNA sequences. *Cytogenet. Cell Genet.* **63:** 97.
Stallings, R.L., D.C. Torney, C.E. Hildebrand, J.L. Longmire, L.L. Deaven, J.H. Jett, N.A. Doggett, and R.K. Moyzis. 1990. Physical mapping of human chromosomes by repetitive sequence fingerprinting. *Proc. Natl. Acad. Sci.* **87:** 6218.
Stallings, R.L., N.A. Doggett, D. Callen, S. Apostolou, L.Z. Chen, J.K. Nancarrow, S.A. Whitmore, P. Harris, H. Michison, M. Breuning, J. Sarich, L. Fickett, M. Cinkosky, D.C. Torney, C.E. Hildebrand, and R.K. Moyzis. 1992b. Evaluation of a cosmid contig physical map of human chromosome 16. *Genomics* **13:** 1031.
Sutherland, G.R. 1988. Fragile sites and cancer breakpoints: The pessimistic view. *Cancer Genet. Cytogenet.* **31:** 5.
Sutherland, G.R. and R.N. Simmers. 1988. No statistical association between common fragile sites and nonrandom chromosome breakpoints in cancer cells. *Cancer Genet. Cytogenet.* **31:** 9.
Tomilin, N.V., V.M. Bozhkov, E.M. Bradbury, and C.W. Schmid. 1992. Differential binding of human nuclear proteins to Alu subfamilies. *Nucleic Acids Res.* **20:** 2941.
van der Feltz, M., M.K. Shivji, P.B. Allen, N. Heisterkamp, J. Groffen, and L.M. Wiedemann. 1989. Nucleotide sequence of both reciprocal translocation junction regions in a patient with Ph positive acute lymphoblastic leukemia, with a breakpoint in the first intron of the BCR gene. *Nucleic Acids Res.* **17:** 1
Vassart, G., M. Georges, R. Monsieur, H. Brocas, A.S. Lequarre, and D. Christophe. 1987. A sequence in M13 phage detects hypervariable minisatellites in human and animal DNA. *Science* **235:** 683.
Wahls, W.P., L.J. Wallace, and P.D. Moore. 1990. Hypervariable minisatellite

DNA is a hotspot for homologous recombination in human cells. *Cell* **60**: 95.
Wallace, M.R., L.B. Andersen, A.M. Saulino, P.E. Gregory, T.W. Glover, and F.S. Collins. 1991. A de novo Alu insertion results in neurofibromatosis type 1. *Nature* **353**: 864.
Wells, R.A., G.G. Germino, S. Krishna, V.J. Buckle, and S.T. Reeders. 1990. Telomere-related sequences at interstitial sites in the human genome. *Genomics* **8**: 699.
Wessels, J.W., H.G. Dauwerse, M.H. Breuning, and G.C. Beverstock. 1991a. Inversion 16 and translocation (16;16) in ANLL M4eo break in the same subregion of the short arm of chromosome 16. *Cancer Genet. Cytogenet.* **57**: 225.
Wessels, J.W., P. Mollevanger, J.G. Dauwerse, F.H. Cluitmans, M.H. Breuning, and G.C. Beverstock. 1991b. Two distinct loci on the short arm of chromosome 16 are involved in myeloid leukemia. *Blood* **77**: 1555.
Wolff, R.K., R. Plaetke, A.J. Jeffreys, and R. White. 1989. Unequal crossingover between homologous chromosomes is not the major mechanism involved in the generation of new alleles at VNTR loci. *Genomics* **5**: 382.
Wong, Z., N.J. Royle, and A.J. Jeffreys. 1990. A novel human DNA polymorphism resulting from transfer of DNA from chromosome 6 to chromosome 16. *Genomics* **7**: 222.

Genetic Control of Simple Sequence Stability in Yeast

Arthur J. Lustig[1] and Thomas D. Petes[2]

[1]Molecular Biology Program, Memorial Sloan-Kettering Cancer Center
and Graduate Program in Molecular Biology, Cornell University
Graduate School of Medical Sciences, New York, New York 10021

[2]Department of Biology
University of North Carolina
Chapel Hill, North Carolina 27599-3280

Simple sequence tracts have been identified in a wide variety of eukaryotic genomes. Such tracts fall into two broad classes: dispersed simple sequence tracts and highly localized tracts associated with specific chromosomal substructures, such as the telomere. Although the function of internal dispersed tracts remains unknown, changes in their tract length are likely to provide a significant source of genomic instability. In contrast, the simple sequences found at telomeres are essential for maintaining the stability of the eukaryotic genome, with the loss of part or all of the telomeric tract leading to cell death. The budding yeast *Saccharomyces cerevisiae* has served as an excellent model system to investigate the stability of both classes of repeats. In this review, we focus on two types of repeats present in the yeast genome: poly(GT) tracts, located at internal sites, and the telomeric poly(G_{1-3}T) tracts. In particular, we emphasize the genetic control of the stability of these repeats and the implication of these studies on the mechanisms involved in altering tract length, which are likely to be shared among diverse eukaryotes.

Specific topics discussed include:

❑ the general structure and stability of eukaryotic dispersed and telomeric simple sequence repeats

❑ the stability of simple sequences in *Escherichia coli*

❑ the processes that act on internal poly(GT) and poly(G_{1-3}T) tracts in yeast

❏ the factors defined genetically and biochemically that act to regulate yeast telomere tract size

INTRODUCTION

The presence of simple sequence repeats within eukaryotic genomes has been documented for more than a decade. Two general classes of simple sequence repeats have been identified. The first class, dispersed simple repeats, are composed of tracts, present at a wide variety of genomic loci, in which a single base (e.g., poly(A)) or a short (2-5 bp) combination of bases (e.g., poly(GT)) is reiterated (Hamada et al. 1982; Lustig and Petes 1984; Tautz et al. 1986). Functions of dispersed simple repeats are not known, but these sequences may well constitute a major source of genomic instability. In particular, dispersed tracts of simple repeats are capable of producing two types of genomic instability. First, recombination between tracts located at different chromosomal positions could cause deletions, inversions, translocations, or other types of chromosomal aberration (for review, see Petes and Hill 1988). A second type of instability, which is the focus of this review, arises through variation in the length of the simple sequence tract. The importance of controlling simple sequence tract length is underscored by recent findings suggesting that expansion and contraction of simple sequences may be causally related to several heritable disease states (Caskey et al. 1992; Huntington's Disease Collaborative Research Group 1993).

The second class of simple sequence repeats consists of the GT-rich tracts found at most eukaryotic telomeres. The maintenance of telomeric simple sequences is essential for the complete replication and stability of linear chromosomes (for review, see Zakian 1989). In fact, loss of telomeric sequences results in high levels of chromosome loss and cell death (Lundblad and Szostak 1989). It has become increasingly evident that novel mechanisms act at the telomere to maintain telomere tract size, but the relationship between processes that act at internal and telomeric tracts remains unknown.

The yeast S. cerevisiae serves as an ideal system to investigate the processes that regulate the size of internal and telomeric simple sequence tracts. Although it has many of the features of dispersed and telomeric repeated elements found in higher eukaryotic genomes, yeast is uniquely amenable to both genetic and molecular approaches to study this problem. The ability to define mutations influencing the stability of yeast simple sequences both facilitates an understanding of the mechanisms regulating tract stability and permits the identification of the genes and gene products important for these processes. The characterization of such genes in yeast should also facilitate the identification of similar genes in other eukaryotes.

In this review, we focus on studies in yeast relating to the regulation of internal poly(GT) and telomeric poly($G_{1-3}T$) tract length. Before describing the yeast studies, however, we review information about simple repetitive DNA sequences in other eukaryotes and in *E. coli*.

SIMPLE SEQUENCE REPEATS IN EUKARYOTES

Dispersed simple sequence repeats

Tracts of DNA (usually 25–60 bp in length) in which a single base or a pair of bases is repeated are common in the eukaryotic, but not the prokaryotic, genome (Hamada et al. 1982; Tautz et al. 1986); poly(A), poly(GA), and poly(GT) tracts are particularly common. There are about 10^5 and 10^4 copies of poly(GT) and poly(A), respectively, in the human genome (Hamada et al. 1982; Lustig and Petes 1984). In addition to homopolymeric and copolymeric tracts, tracts involving duplications of 3–5 bp (microsatellites) are also frequent in the mammalian genome (for review, see Jeffreys et al. 1991). Although tracts of simple repeats tend to be dispersed throughout the genome, in some organisms, simple repeats are arranged in very long tandem arrays associated with specific chromosomal structures, as observed in the satellite sequences of *Drosophila* (Lohe and Brutlag 1987).

It is not clear whether dispersed simple repeats serve any useful function in the cell. Most of the available information concerns poly(GT) tracts. Lowenhapt et al. (1989) found that poly(GT) tracts are nonrandomly dispersed on *Drosophila* chromosomes in patterns suggesting their involvement in the regulation of transcription and/or recombination. Direct tests of the effect of poly(GT) on transcription have yielded differing results, exerting both positive and negative effects in different systems (Berg et al. 1990; Naylor and Clark 1990). Poly(GT) tracts appear to stimulate recombination in both mammalian cells (Stringer 1985; Wahls et al. 1990) and yeast (Treco and Arnheim 1986). Although these studies indicate that poly(GT) tracts affect biologically important processes, these experiments do not indicate that these sequences have an essential role in the eukaryotic genome. The function of poly(A) tracts in higher eukaryotes is similarly unclear. However, the majority of these tracts are present at the 3′ end of *Alu* elements, suggesting that they were introduced during *Alu* transposition (Lustig and Petes 1984).

As noted above, one class of genomic instability elicited by dispersed simple sequences is a change in the size of the tract. Alterations in tract length are important for a number of reasons. First, for four different diseases (fragile X syndrome, myotonic dystrophy, Kennedy's disease, and Huntington's disease), expansions of tracts of simple repeats appear causally related to the disease (Caskey et al. 1992; Huntington Disease Collaborative Research Group 1993). Second, since poly(GT) tracts are frequently used as polymorphic genetic markers (Weber and May 1989), it is important to know the rate of de novo length alterations. In one human population study, no alterations for one particular poly(GT) tract were observed in 176 individuals examined (Weber and May 1989).

Two mechanisms have been proposed to cause instability of simple repetitive DNA sequences: unequal recombination (Fig. 1a) and DNA polymerase slippage (Fig. 1b). Unequal recombination involves base-pairing between allelic tandem arrays that are misaligned by an integral number of repeats; the interacting arrays could be located on homologous chromosomes or on sister strands of a single chromosome. Crossing-over between such arrays produces one tandem array with more repeats and another array with fewer repeats. Alternatively, a gene-conversion event may alter the number of repeats in a nonreciprocal manner (Fogel et al. 1984). An alternative mechanism causing changes in tract length is DNA polymerase slippage (Streisinger et al. 1966). In this mechanism (Fig. 1b), there is a transient dissociation of template

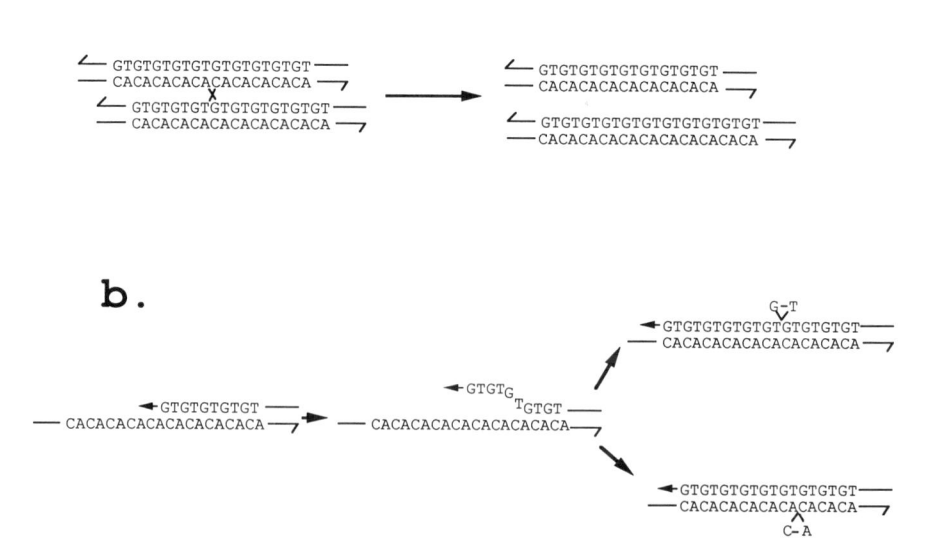

Figure 1 Mechanisms that alter the length of simple repetitive tracts of DNA. (a) Unequal crossing-over. Tracts from allelic positions on two DNA molecules pair in a misaligned configuration. Crossing-over generates one tandem array with an addition of one GT and one with the reciprocal deletion. (b) DNA polymerase slippage. During DNA replication, the primer and template strands transiently dissociate. If the primer reassociates with the template with unpaired bases in the primer strand, continued DNA synthesis creates a longer tract (upper portion of figure). If the reassociation results in unpaired bases in the template strand, continued synthesis results in a shorter tract (lower portion of figure).

and primer strands during replication of the tract. If the strands reanneal in a misaligned configuration, continued synthesis could lead to altered tract lengths. If the misalignment produces a bulge in the primer strand, one of the tracts will be longer than the original tract. Alternatively, if the misalignment results in a bulge in the template strand, the altered tract will be shorter.

Although the stability of simple repeats during in vitro DNA replication has not been examined explicitly, Kunkel (1986) showed that eukaryotic DNA polymerases frequently cause frameshift mutations during in vitro replication of short polymeric tracts; these mutations are almost always small deletions.

Telomeric simple sequences

Structure of the eukaryotic telomere A second class of eukaryotic simple sequences is that associated with chromosomal telomeres. In the majority of eukaryotes, chromosomes terminate in tandem arrays of 2-bp to 16-bp repeating units following the consensus $[(T/A)_{1-4}N_{0-4}G_{1-8}]$ (for review, see Zakian 1989). These arrays consist either of exact repetitions of a discrete unit or of an irregular repeating unit. In these telomeres, the repeating unit contains an asymmetrical distribution of G and C residues, with the G-rich strand invariably proceeding 5' to 3' toward the telomere. Tract length varies in different organisms from 36 bp in *Oxytricha* (Klobutcher et al. 1981) to more than 20 kb in some species of mice (Kipling and Cooke 1990). In the several systems examined to date, the extreme end terminates in a 3' overhang (Klobutcher et al. 1981; Henderson and Blackburn 1989; Wellinger et al. 1993), which in vitro is capable of forming intramolecular and intermolecular structures through G-G base-pairing (Henderson et al. 1987; Sundquist and Klug 1989; Williamson et al. 1989; Panyutin et al. 1990; Sen and Gilbert 1990). However, exceptions to this strand asymmetry have recently been reported, raising doubts as to the functional significance of these structures (Muller et al. 1991; Teschke et al. 1991; McEachern and Hicks 1993). Centromere-proximal to the simple sequence elements are longer species-specific, and often end-specific, repetitive elements. Interstitial telomeric tracts are often associated with these elements (Walmsley et al. 1984; Weber et al. 1990; Wells et al. 1990; Louis and Haber 1992).

Control of telomere tract size In most organisms, tract addition is an inexact process. The tract size of even an individual telomere varies among cells of the population. Nonetheless, the process is regulated, since tract sizes are normally clustered within a discrete distribution centered around a constant average tract length. Two sets of processes act at the telomere to regulate tract size. The first set involves factors that elongate telomeric DNA. As first noted by Watson (1972) and Olkovnikov (1971), in the absence of a compensatory mechanism, degradation of the terminal RNA primer results in loss of terminal sequences on one of the two sister chromatids following DNA replication. Continual deletion through repetitive cycles of DNA replication would be expected to lead to cell death as the consequence of either loss of an essential gene or the inability of the truncated telomere to perform an essential function. Greider and Blackburn (1985) identified a ribonucleoprotein, telomerase, that is likely to be responsible for telomere tract addition in many organisms. This enzyme, initially isolated from *Tetrahymena* macronuclei, is capable of addition of telomeric repeats onto the G-rich strand of single-stranded telomeric substrates. Interestingly, the RNA component of telomerase acts as template for telomere addition both in vitro and in vivo, suggesting that telomerase may act as a specialized reverse transcriptase (Shippen-Lentz and Blackburn 1990; Yu et al. 1990). The identification of telomerase in human cells (Morin 1989) suggests that it is a widely used mechanism for telomere synthesis. However, telomerase-based systems may not be universal. Indeed, some dipteran species appear to have lost telomere-like simple sequence tracts altogether (Biessmann et al. 1990; Okazaki et al. 1993) and may have supplanted this mechanism with telomere-specific transposition of longer repeated elements (Biessmann et al. 1992).

The maintenance of an equilibrium centered around an average tract length suggests the presence of a "ruler" mechanism that acts to compensate for telomere elongation (Lustig and Petes 1986). The properties of telomerase do not easily explain this phenomenon, since telomerase, at least in vitro, acts processively to produce extremely long telomeres (Greider 1991). It is therefore likely that other factors act to remove telomeric tract sequences or to limit tract addition in a size-dependent fashion.

In some organisms, physiological conditions can affect telomere length. *Tetrahymena* macronuclear telomeric tracts slowly increase in size during logarithmic growth but spontaneously lose much of this increase in length during stationary phase (Larson and Blackburn 1987). A similar telomere elongation process has been observed in trypanosome telomeres following multiple passages in host organisms (Bernards et al. 1983). These elongated telomeres appear to be unstable and undergo deletion during cloning or after heat shock. In the yeast *Candida*, telomere tract size is temperature-sensitive, with higher temperatures increasing the number of 23-bp repeats present at these telomeres (McEachern and Hicks 1993). The functional significance of these physiological changes in tract length remains unknown.

In contrast to the controlled accessibility of the telomere to factors that maintain telomere tract length, some recombinational (and nucleolytic) factors are likely to be excluded from the telomere. Early studies initiated by Barbara McClintock (1941) demonstrated that internal chromosomal breaks fuse randomly in maize to produce unstable dicentric chromosomes. In contrast, telomeres never participate in such reactions in normal cells. In addition, in yeast, linear plasmids terminated in random sequence are subject to both degradation and end-to-end fusion, whereas plasmids terminated in telomeric sequence appear to be resistant to these processes (Kunes et al. 1990; Lustig et al. 1990; Lustig 1992). These characteristics may be the consequence of the unusual structure of telomeric chromatin. Nonnucleosomal telomeric chromatin has been identified in both ciliates and yeast (Gottschling and Cech 1984; Budarf and Blackburn 1986; Price 1990; Wright et al. 1992). In the macronuclei of *Oxytricha* and *Euplotes*, telomeric chromatin consists of two proteins organized in heterodimeric complexes that associate with single-stranded 3'overhangs (Gottschling and Zakian 1986; Price 1990). Interestingly, chromatin reconstituted in vitro demonstrates resistance to nuclease *Bal*31, suggesting a similar protective function in vivo (Gottschling and Zakian 1986). A protein with similar properties has recently been identified in *Xenopus* eggs (Cardenas et al. 1993). The ability of single-stranded G-rich DNA to form intramolecular G-G basepaired structures could conceivably also afford protection against some nucleolytic and recombinational processes.

Telomere tract size and cellular senescence Several lines of evidence suggest a link between telomere tract size and cellular senescence. First, telomere sizes in human germ-line cells are consistently larger than those observed for somatic tissues (Allshire et al. 1989; Cross et al. 1989; deLange et al. 1990). Second, during senescence of cultured human fibroblasts and kidney cells, telomere tract size continually decreases until cell proliferation ceases (Harley et al. 1990; Counter et al. 1992). In contrast, telomere size of immortalized cells is stably maintained (Counter et al. 1992). The number of doublings prior to senesence strongly correlates with the initial size of the telomere tract from the donor cells (Allsopp et al. 1992). Cells containing longer telomeres are capable of additional rounds of replication before cell division arrest. Third, telomere tracts are shorter in tumor cells compared with normal cells from the same tissue (Hastie et al. 1990), as expected if additional doublings result in telomere tract loss in vivo. Fourth, patients suffering from Hutchinson-Gilford progeria, a disease causing premature aging, have significantly shorter telomeric tracts, and cells from these patients, when cultured in vitro, have a reduced replicative capacity (Allsopp et al. 1992). Finally, senescing cells lack measurable telomerase activity, whereas immortalized cell lines and germ-line cells contain high levels of telomerase activity (Counter et al. 1992). A decrease in the mean tract length or the specific loss of a single telomere could trigger senescence through loss of, or alteration in, the transcription of an essential gene;

activation of a cell-cycle checkpoint; or formation of unstable dicentric chromosomes.

Although the correlation between telomere length and cell line senescence is compelling, a causal relationship remains to be established. Alternative interpretations in which the decrease in tract length and loss of telomerase activity are secondary consequences of the pleiotropic effects of senescing cells (or disease states) remain a serious possibility. That tract loss is involved in normal aging processes is far less clear. Indeed, the correlation between telomere length and age in humans is rather weak (Allsopp et al. 1992) and, at least in one species of mice, mean length does not change with age (Kipling and Cooke 1990).

STABILITY OF SIMPLE REPEATS IN *E. COLI*

Although the stability of long (>20 bp) simple repeats has only recently been examined in *E. coli*, many studies of spontaneous mutations in *E. coli* indicate that small direct repeats are hot spots for deletions, and short (3-5 bp) homopolymeric tracts are mutational hot spots (for review, see Ripley 1990). In some studies (see, e.g., Timmons et al. 1986), the frequency of mutations involving small repeats is *RecA*-independent, implicating DNA polymerase slippage as the primary mechanism. Alternatively, recombination between small repeats or recombination within a repetitive tract might represent a *RecA*-independent type of recombination. The hypothesis that in vivo DNA polymerase slippage causes alterations in tract length is supported by the observation that *E. coli* DNA polymerase slips in vitro on simple sequence templates (Kornberg et al. 1964; Schlotterer and Tautz 1992).

The instability of long simple sequence tracts in *E. coli* was investigated by Levinson and Gutman (1987) and Freund et al. (1989) using similar procedures. Plasmids were constructed in which the repetitive tract was inserted into the coding sequence of β-galactosidase. In some experiments, the insertion was in-frame and, therefore, alterations in tract length that resulted in loss of β-galactosidase activity could be detected using X-gal medium. In other experiments, the insertion was out-of-frame, and X-gal plates were used to detect variants that restored the correct reading frame. A 40-bp tract of poly(GT) cloned in an M13 vector altered length at a rate of about 1% (Levinson and Gutman 1987). Most of the alterations were small (2 or 4 bp) deletions, although a few small insertions were also observed. Mutations in genes involved in mismatch repair (*mutL* and *mutS*) resulted in tenfold elevated levels of tract instability, whereas mutations in *recA* had no obvious effect on stability. Using a similar system, Freund et al. (1989) showed that alterations in a 48-bp poly(GT) tract occur at a rate of 6×10^{-3}. Most of these changes were large deletions between 10 bp and 22 bp in length.

In summary, a number of generalizations can be made concerning the behavior of poly(GT) tracts in *E. coli*. First, the tracts are quite unstable; the stability in single-stranded vectors is somewhat less than in double-stranded vectors. Second, the stability is unaffected by *recA* mutations but is tenfold reduced by mutations in the mismatch repair systems. These effects are consistent with those expected if the alterations in tract length reflect DNA polymerase slippage. Third, in at least some contexts, the tracts tend to shrink in length (Freund et al. 1989).

GENETIC CONTROL OF THE STABILITY OF SIMPLE REPETITIVE DNA SEQUENCES IN YEAST

There has been no systematic study of simple repetitive DNA sequences in *S. cerevisiae*. Southern analysis of yeast DNA indicated that there are about 100 sequences that hybridize to poly(GT) (Hamada et al. 1982; Walmsley et al. 1983). Although many of these sequences were subsequently shown to represent tracts of poly(G_{1-3}T) located at or near the telomeres (Walmsley et al. 1983, 1984; Shampay et al. 1984), one non-telomeric tract contained alternating GT sequences, 34 bp in length, similar to those observed in higher eukaryotes. As in higher eukaryotes, the function (if any) of these simple repeats in yeast is unclear. Although, in some contexts, poly(GT) (Treco and Arnheim 1986) and poly(G_{1-3}T) (M. White et al., unpubl.) tracts appear to stimulate meiotic recombination, it is not clear that this stimulation is their primary role. Regardless of their primary function, simple repeats represent potential sources of instability in the yeast genome.

Experimental systems to probe simple sequence tract instability

The stability of these tracts in yeast was examined using experimental systems closely related to that developed by Levinson and Guttman (1987) and discussed above (Henderson and Petes 1992). One plasmid detection system involved the insertion of a poly(GT) tract 29 bp in length into the coding sequence of a β-galactosidase gene (derived from *E. coli*) that was fused to the promoter of the yeast *LEU2* gene, a promoter that functions in both yeast and *E. coli* (Tu and Casadaban 1990). This plasmid (Fig. 2a) also contained a selectable yeast gene (*TRP1*), replication origins for both *E. coli* and yeast, a selectable drug resistance (*amp^R*), and a centromere. The centromere prevents loss of the plasmid and ensures that yeast cells have only one or two copies of the plasmid. When yeast cells were transformed with this plasmid, the transformed cells formed white colonies on the chromogenic substrate X-gal because the insertion of poly(GT) caused a frameshift mutation. At a rate of 10^{-4}, derivatives that formed blue colonies (indicating synthesis

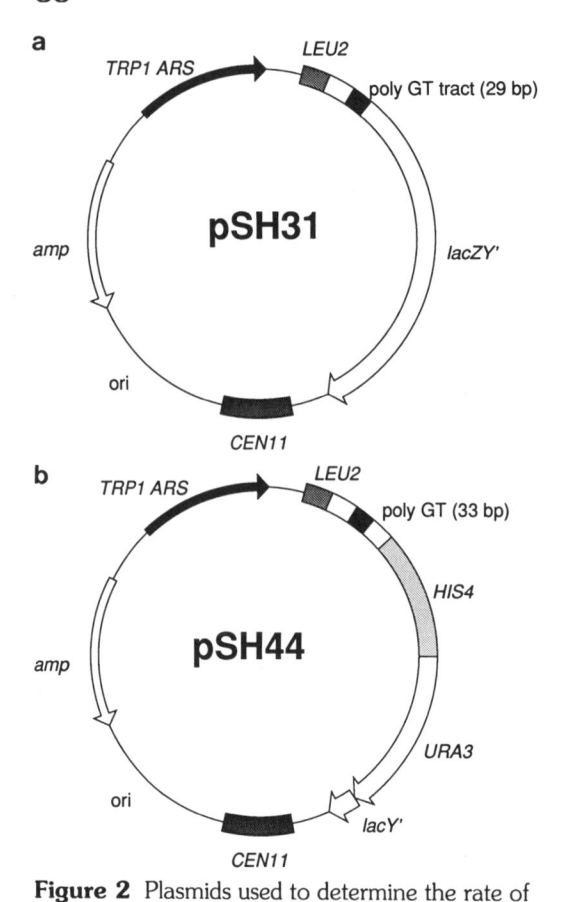

Figure 2 Plasmids used to determine the rate of instability of poly(GT) tracts in yeast. (a) pSH31. The plasmid contains selectable markers and replication origins that function in S. cerevisiae and E. coli. The β-galactosidase gene is expressed as the result of fusion to the yeast LEU2 promoter (functional in both yeast and E. coli). The poly(GT) insertion (29 bp) is within the coding sequence of β-galactosidase. Since the insertion is out-of-frame, yeast and E. coli cells containing the plasmid form white colonies on X-gal plates. Alterations of the tract length to restore the correct reading frame are detected by the formation of blue colonies on X-gal.
(b) pSH44. In this plasmid, the LEU2 promoter results in transcription of a fusion protein that has a small amount of the β-galactosidase and HIS4 genes, and almost the entire URA3 gene. Since the poly(GT) insertion in this protein is in-frame, cells containing the plasmid are phenotypically Ura$^+$. Alterations in tract length are selected on plates containing 5-fluoro-orotate, which selects for the mutant ura3 genotype.

of functional β-galactosidase) were produced.

Plasmids derived from these cells were rescued into *E. coli* and sequenced. As expected, most of these plasmids had alterations in the length of the poly(GT) tract that restored the correct reading frame. Of 25 tracts examined, most contained deletions of 2 bp (17) or additions of 4 bp (3). Of the remaining 5 tracts, 2 had insertions of 2 bp, 1 had a deletion of 4 bp, 1 had a deletion of 14 bp, and 1 had a deletion of 2 bp and an addition of 15 bp of telomeric poly(G_{1-3}T) DNA. There were two surprises in this study. First, three of the tracts (those with an addition of 2 bp or deletion of 4 bp) were out-of-frame and produced β-galactosidase; the cells containing these plasmids were pale blue on X-gal plates instead of the dark blue characteristic of the in-frame plasmids. The production of β-galactosidase from these cells is likely to reflect translational frameshifting, since poly(GU) tracts in the +1 reading frame have been shown to have a high rate of this process (Weiss et al. 1987). A second unexpected finding was the insertion of telomeric sequences into the poly(GT) tract. It is possible that the poly(GT) sequences can engage in ectopic gene conversion events with telomeric sequences. Alternatively, at low frequency, telomerase may be able to modify poly(GT) tracts into telomeric sequences.

Two other observations were made concerning the stability of the poly(GT) tracts. First, the rate of instability of poly(GT) tracts was not affected by the orientation of the tract with respect to the origin of replication of the plasmid. Second, tracts inserted into the yeast genome also had a high rate of instability (about 0.3×10^{-4}/division). Thus, the instability is not an artifact of the plasmid system. The stability of two other types of tracts was examined, poly(G) (20 bp) and poly(G_{1-3}T) (51 bp). The poly(G) tracts were about twofold less stable than the poly(GT) tracts, whereas the poly(G_{1-3}T) tracts were substantially (about tenfold) more stable.

In considering the origin of repetitive DNA sequences in the eukaryotic genome, one relevant issue is whether alterations in tract length are more often insertions or deletions. In the studies described of the plasmid-borne poly(GT) tract inserted upstream of β-galactosidase, deletions appeared to be more common than insertions (32 deletions and 7 insertions). These numbers, however, may be biased by the system used to detect altered tracts, since a deletion of 2 bp yields a dark blue colony, whereas an addition of 2 bp yields a pale blue colony. To examine the relative frequencies of additions and deletions without this bias, a new experimental system was designed (Fig. 2b). A plasmid (pSH44) was constructed in which an in-frame poly(GT) tract of 33 bp was inserted into a gene encoding a fusion protein containing a small portion of the *HIS4* gene attached to *URA3* (Henderson and Petes 1992). Yeast strains containing this plasmid were phenotypically Ura⁺. Tract alterations resulting in an out-of-frame insertion were selected by plating the cells on 5-fluoro-orotate, which selects for the growth of Ura⁻ cells (Boeke et al. 1984). With this system, there should be no bias in detecting +2 and −2 alterations. Of 22 altered tracts sequenced, 16 contained insertions and 6 contained deletions. Although larger numbers of tracts must be analyzed in this system, these preliminary results suggest that, in at least some contexts, insertions may occur more frequently than deletions.

Genetic control of simple sequence tract stability

One approach to determining the mechanism of tract instability is to investigate the effects of various mutations on the frequency of alterations. The mutation *rad52* reduces the frequency of most, but not all, types of recombination in yeast (for review, see Petes et al. 1991). This mutation has no quantitative or qualitative effect on the stability of poly(GT) tracts (Henderson and Petes 1992). Null mutations in the *RAD5* gene increase tract stability about 10-fold (Johnson et al. 1992). Sequence analysis of *RAD5* suggests that this gene may encode a DNA helicase, although no biochemical studies have been done thus far. Mutations in the *PMS1* gene result in 100-fold decreased levels of tract stability (M. Strand and T. Petes, unpubl.). *PMS1* encodes a protein that is homologous to the *MutL* gene product of *E. coli* (Kramer et al. 1989), a protein that is involved in correction of base-pair mismatches (Williamson et al. 1985).

Several results strongly implicate DNA polymerase slippage as the mechanism leading to tract instability. First, the rate of tract instability (10^{-4}) is very high relative to recombination events involving other types of small repeats in yeast (10^{-10}; Ahn et al. 1988). Second, the rate of tract instability is unaffected by the *rad52* mutation, which reduces the rate of most types of recombination. Third, the instability is greatly increased by mutations in the *pms1* gene. This result is expected if DNA polymerase slippage events frequently produce DNA duplexes with bulges (Fig. 1b). We suggest that the mismatch repair system usually corrects the resulting mismatch by excising DNA from the newly synthesized strand. In the absence of this system, one would expect a great elevation in tract instability. By the mechanism shown in Figure 1b, slippage events are a consequence of dissociation of the newly synthesized strand from the template strand. One way of explaining the observation that *rad5* mutations increase tract stability is to suggest that this dissociation requires the putative *RAD5* helicase activity. It should be pointed out that DNA polymerase slippage in yeast has also been implicated in the deletion of palindromic sequences from the genome (Henderson and Petes 1993; Ruskin and Fink 1993; D. Gordenin et al., pers. comm.).

In summary, these results indicate that the rate of alterations in the length of simple repeats is likely to be affected by a large number of parameters: the initial rate of errors (a property of DNA polymerase and cofactors), the probability of dissociation of template and primer strands (a property of associated helicases), and the probability of correction of the resulting mismatch (a property of mismatch correction systems).

GENETIC CONTROL OF THE STABILITY OF TELOMERIC REPEATS IN YEAST

Telomere structure in *S. cerevisiae*

The second major class of repeated elements in yeast consists of the simple sequences located at the extreme end of the telomere. The simple sequence tract follows the consensus poly($G_{2-3}T(GT)_{1-6}$) (abbreviated poly($G_{1-3}T$)), the G-rich strand oriented 5′ to 3′ toward the telomere (Fig. 3) (for review, see Zakian 1989). The average tract length varies from 150 bp to 800 bp in different laboratory strains. In any given strain, however, the tract is maintained at a constant average length, with the length of an individual telomere differing by only 100 bp within a population (Walmsley and Petes 1985). Centromere-proximal to the telomeric tract are the highly conserved Y′ elements present at many telomeres in one to four tandem copies (Chan and Tye 1983). Further proximal is the less conserved X element present in single copy at most, if not all, telomeres. Both X and Y′ elements contain potential origins of replication, some of which are activated late in S phase (Chan and Tye 1983; Ferguson et al. 1991; Ferguson and Fangman 1992). Poly($G_{1-3}T$) tracts of shorter size (50–150 bp) are found between reiterated Y′ elements and at some XY′ junctions (Walmsley et al. 1984; Louis and Haber 1992; C. Chan and B.-K. Tye, pers. comm.).

Two additional classes of telomere-associated simple sequence repeats have been identified. First, Horowitz and Haber (1984) identified

Figure 3 Structure of the yeast telomere. The yeast telomere terminates in poly($G_{1-3}T$) tracts (*light-striped box*). Centromere-proximal to these repeats are two additional long repeated elements, the 5.2–6.7-kb Y′ element (*dark-striped box*), present in 1–4 copies at many telomeres, and the variably sized X element (*stippled box*). Additional poly($G_{1-3}T$) tracts are often present between reiterated Y′ elements and at the XY′ junction. Both Y′ and X contain autonomously replicating sequence (ARS) elements, some of which have been demonstrated to function as late-activating origins of replication in vivo. The 36-bp repeats within the Y′ element are shown by the vertical lines.

tandemly arrayed 36-bp imperfect repeats within each Y' element. The number of these repeats ranges from 8 to 25 among Y' elements from different strains and from different telomeres of the same strain, suggesting that they may be the site of frequent unequal sister chromatid exchange or gene conversion. Interestingly, the 36-bp repeats often contain several GT-rich telomere-like sequences (GTTGGT and TGTTGG). Second, several copies of the human telomeric repeat T_2AG_3 are found at junctions both between the X and Y' elements and between the X element and poly($G_{1-3}T$) tracts and are often associated with neighboring degenerate repeats within the X element (Liu and Tye 1991; Louis and Haber 1991). Three to four copies are also clustered within the Y' element 100 bp from the telomeric tract (Louis and Haber 1992). Whether these GT-rich sequences serve a function or are originally derived from telomeric sequences is unknown.

The poly($G_{1-3}T$) tracts have several unusual structural features. First, these tracts appear to contain multiple nicks and gaps, a property that is shared with other eukaryotic telomeres (Szostak and Blackburn 1982). Second, the poly($G_{1-3}T$) tract terminates in a single-stranded 3' overhang of more than 50 bp in length (Wellinger et al. 1993). Interestingly, the 3' overhang appears only transiently in late S phase, after which it is rapidly eliminated. These data suggest that the overhang may represent an intermediate in telomere replication, formed immediately following telomere addition or RNA primer (or exonucleolytic) loss. Third, single-stranded telomeric sequences in yeast, as in other organisms, form highly folded intramolecular structures in vitro through G-G Hoogsteen base-pairing (Henderson et al. 1987; Lustig 1992). The observation that plasmid telomeres incapable of forming these structures nonetheless serve as efficient substrates for poly($G_{1-3}T$) addition in vivo (Lustig 1992) argues against an essential role for these structures in telomere addition. Fourth, yeast telomeric tracts are packaged in vivo into nonnucleosomal chromatin structures (Wright et al. 1992). These structures are resistant to micrococcal nuclease and DNase I in vitro and have a size roughly equivalent to the length of the telomeric tract.

Telomere size control in wild-type cells

In any given strain of yeast, the telomere tract is maintained at a constant average size, with individual telomeres appearing randomly at sizes greater than or less than the average. In part, this equilibrium between the loss and gain of telomeric sequence is a stochastic process. Shampay and Blackburn (1988) demonstrated that wild-type cells randomly inheriting an individual elongated telomere maintain that telomere at elongated sizes after further growth, indicating that the size of a tract is governed in part by the length of the inherited telomere. Nonetheless, they also found limits to the maximal size that a telomere could attain in any given strain, suggesting the presence of additional factors that regulate this process. The maintenance of this equilibrium is both *rad1*- and *rad52*-independent, indicating that the major pathways of mitotic recombination are not involved in this process (Shampay and Blackburn 1988; A. Lustig, unpubl.).

Wang and Zakian (1990) demonstrated that telomere loss and addition are restricted to the distal portion of the tract. Sequencing of linear plasmid telomeres, after transformation into yeast, revealed that the internal 120–150 bp of the 300-bp telomeric tract is conserved among subcultured progeny of a given transformant, whereas the terminal 150 bp is subject to rapid variations in sequence. These data suggest either that the mechanisms of telomere loss (nucleolytic degradation and primer loss) rarely extend beyond the terminal 150 bp before telomere addition occurs or, alternatively, that the telomere consists of distinct domains differing in resistance to degradation.

Genetic analysis of telomere tract length regulation in yeast

That telomere tract size is controlled by a discrete number of gene products was first demonstrated by Walmsley and Petes (1985), who examined the meiotic progeny of crosses between two nonisogenic laboratory strains that differed in mean telomere tract size. The telomeres of different spore colonies attained several distinct mean sizes affecting all telomeres uniformly, suggesting the presence of multiple gene products that differed between these two strains.

Several genetic approaches have subsequently been taken to identify individual components involved in telomere tract size control. Lustig and Petes (1986) screened directly for recessive mutations that resulted in a slow loss of telomeric sequences. Such a phenotypic lag is expected if a small amount of primer loss and exonucleolytic degradation occurs during each cell division in the absence of terminal addition. Two complementation groups, *TEL1* and *TEL2*, were identified in this screen. *tel1* and *tel2* mutations result in a slow decrease of tract size from 360 bp present in the isogenic parent strain to 70 bp and 150 bp, respectively. Further subculturing does not result in additional decreases in tract size. Since there is no selective growth disadvantage conferred to cells having the shorter telomeres, these data suggest that a new equilibrium was reached that balanced a reduced efficiency of terminal addition with a reduction in the loss of telomeric sequences. This result implies that the efficiency of either telomere addition or telomere loss is sensitive to tract length.

A second approach was taken by Lundblad and Szostak (1989), who identified a mutation, *est1*, that fails to resolve circular plasmids containing telomeric inverted repeats into linear forms. *est1* mutant cells exhibit a slow, but continual, loss of telomeric sequences. This loss of sequences is correlated with an increase in the frequency of both chromosome loss and cell death. It is unclear whether cell death is the consequence of loss of an essential gene after telomere degradation, loss of an essential telomeric function after a minimal size is attained, or arrest by a cell-cycle checkpoint that measures telomere tract length. Although the characteristics of *tel1*, *tel2*, and *est1* mutations are consistent with their direct involvement in telomere addition, the proteins encoded by these genes remain uncharacterized.

Several mutations in genes encoding components of the DNA replication apparatus affect telomere tract size. The best studied of these are mutations in the *CDC17* gene, encoding the catalytic subunit of yeast DNA polymerase I (Carson and Hartwell 1985). Several recessive temperature-sensitive alleles of *cdc17* cause a slow increase in telomere tract size under either permissive or semi-permissive conditions, and telomeres attain sizes up to 2 kb. These phenotypes are not the consequence of a *rad52*-dependent recombinational pathway for telomere elongation, since *cdc17* and *rad52cdc17* double mutants have identical telomere size phenotypes. Mutations in *CDC2*, encoding yeast DNA polymerase III, also result in an increase in telomere tract length (Carson and Hartwell, pers. comm.; A. Lustig, unpubl.). The cause of telomere tract expansion in these mutants remains unexplained. Conceivably, the increased telomere size may be the consequence of an increase in nicks and gaps in the telomere that may provide a substrate for *rad52*-independent recombination pathway. Alternatively, mutant DNA polymerases may result in net elongation following slippage during replication. Tract elongation could also be a consequence of physical or temporal associations between the DNA polymerases and components of the telomere addition machinery. Mutations in two additional cell-cycle genes, *CDC8*, encoding thymidine kinase, and *CDC21*, encoding thymidylate synthetase, cause a slight 50-bp decrease in tract size (S. Kronmal and T.D. Petes, unpubl.). The mechanism of these proteins in telomere addition is unknown, but they may act simply through decreasing thymidine pool size, possibly reducing the efficiency of a yeast telomerase.

Role of telomere-binding proteins in telomere stability

RAP1 Several telomere-binding proteins have been characterized in yeast. The most abundant of these is RAP1 (repressor/activator protein 1). RAP1 was identified in numerous laboratories as a protein capable of binding in vitro to sequences from a wide variety of genomic loci including upstream activation sequences, the silencer elements flanking the cryptic mating-type information at *HML* and *HMR*, and sites embedded within the poly($G_{1-3}T$) tracts of telomeres (for references, see Kyrion et al. 1992). RAP1 binds duplex DNA having the consensus sequence

(G/A)(G/A)TGN(G/A)(C/T)GG(G/A)T(G/T)(C/T) (Buchmann et al. 1988; Vignais et al. 1990). Indeed, the highest affinity binding sites are the GGTGTGTGGGTGT sequences found within telomeric tracts at an average frequency of 1 in 35 bp (Wang and Zakian 1990). Genetic evidence has demonstrated that RAP1 plays a role at each class of sites (Lustig et al. 1990; Kurtz and Shore 1991; Sussel and Shore 1991; Kyrion et al. 1992, 1993). Although RAP1 is essential for viability, the nature of the essential function remains unknown (Shore and Nasmyth 1987). In this review, we focus only on genetic and molecular studies relating to the function of RAP1 at the telomere. That RAP1 is present at the telomere is indicated both by the association of RAP1 with nonnucleosomal telomeric chromatin fractions (Wright et al. 1992) and by immunocytochemical localization of RAP1 to telomeric regions in meiotic cells (Klein et al. 1992).

Several lines of genetic evidence suggest a functional role for RAP1 at the telomere. First, Lustig et al. (1990) and Conrad et al. (1990) characterized several recessive temperature-sensitive missense mutations in the *RAP1* gene. These mutations result in a slow temperature-dependent loss of 50–75% of the poly($G_{1-3}T$) sequences. This phenotype is the likely consequence of reduced binding of RAP1 to DNA at elevated temperatures, as suggested by the temperature sensitivity of Rap1^{ts} protein binding in vitro (Kurtz and Shore 1991). Second, binding of RAP1 to the telomeric tract appears to be important for telomere addition. Mutations introduced within the RAP1-binding sites of synthetic telomeres that reduced or eliminated RAP1 binding in vitro significantly reduced the ability of these telomeres to serve as substrate for poly($G_{1-3}T$) addition in a plasmid telomere healing assay (Lustig et al. 1990).

The study of several additional *rap1* alleles suggests a complex role of RAP1 in the regulation of telomere size. These alleles (*rap1-17*, *rap1-18*, and *rap1-19*, collectively termed the *rap1^t* alleles) have extreme effects on telomere size and stability (Kyrion et al. 1992). First, each of these mutations results in promiscuous telomere addition. Telomere tracts attain sizes up to 4 kb longer than the 300-bp tracts present in wild-type cells. This phenotype is semi-dominant, suggesting an alteration, rather than loss, of RAP1 function. This effect is specific to telomeric poly($G_{1-3}T$) tracts, since an 80-bp internal tract introduced into these mutants does not show a change in tract length. Second, telomere tract size is highly unstable. Although individual telomeres in wild-type cells do not vary by more than 100 bp in size, telomeres of *rap1^t* cells can differ by more than 2 kb after only 25 generations of growth. This heterogeneity is not the consequence of increased telomere size per se, since wild-type cells inheriting elongated telomeres from heterozygous diploids display wild-type size heterogeneity. A second indication of tract instability is the frequent *rad1*- and *rad52*-independent loss of part or all of the increased tract length in a single cell division.

These deleted forms arise in 10–20% of *rap1-17* and *rap1-18* colonies after 25 generations of growth. These rapid deletion events are unlike the phenotypes of most mutations affecting telomere tract size, which exhibit only slow changes in tract length. Third, sequencing of telomeres from *rap1-17* cells has revealed alterations from consensus motifs, including the presence of multiple A and C residues and noncanonical arrangements of G and T residues (K. Boakye and A. Lustig, unpubl.). Interestingly, deviations from the consensus sequence and sequence differences among *rap1-17* telomeres in the population extend to within 65 bp of the beginning of the 1.5-kb poly($G_{1-3}T$) tract. These data indicate that a substantial portion of the telomere tract is rearranged in these cells after only limited growth and are consistent with a loss of fidelity in either a DNA-repair or telomere-addition process.

Each *rap1^t* allele is the consequence of a nonsense mutation resulting in a truncation of the carboxy-terminal 144–165 amino acids of RAP1. These truncated proteins nonetheless bind DNA in vitro efficiently and specifically. These data suggest that the carboxyl terminus of RAP1, although dispensable for DNA binding and viability, is critical for the regulation of telomere size, stability, and fidelity. Since it is unlikely that RAP1 directly associates with factors involved in each process, we favor the notion that an altered structure of telomeric chromatin in *rap1^t* cells exposes the poly($G_{1-3}T$) tract to a multiplicity of factors that are normally more tightly regulated in wild-type cells. It is additionally possible that RAP1 mediates some of its functions by targeting telomeres to specific nuclear substructures. Consistent with this, *rap1^t* cells are defective in the exclusive localization of RAP1 to the nuclear periphery that is observed in wild-type cells (Klein et al. 1992; A. Lustig and S. Gasser, unpubl.).

In vitro mutagenesis of the carboxyl terminus of RAP1 has been used to further localize two specific regions important for the regulation of telomere size (Fig. 4). The first, defined by Sussel and Shore (1991), maps to amino acids 726, 727, and 747 (defined as the *rap1^s* alleles, Fig. 4). Mutations in this region cause a 50-bp to 250-bp increase in mean tract size. The second region maps to the carboxy-terminal 25 amino acids of RAP1 (C. Liu and A. Lustig, unpubl.). Truncation of this carboxy-terminal tail results in both a 450-bp increase in tract size and an increase in tract heterogeneity. The effects of these mutations on the fidelity of telomere addition have not been investigated.

The phenotypes of RAP1 carboxy-terminal mutations are likely to reflect the interactions between the RAP1 carboxyl terminus and other factors important for telomere size regulation. Consistent with this, Conrad et al. (1990) have demonstrated that overproduction of the carboxyl terminus of RAP1 also causes an increase in telomere tract size, suggesting the titration of a specific factor. One such factor, RIF1 (RAP1 interacting factor) has been identified by Hardy et al. (1992b). Mutations in

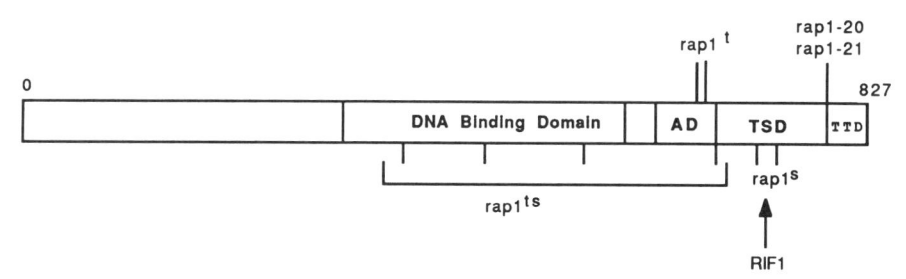

Figure 4 Domain structure of RAP1. A tentative map of the domain structure of the 827-amino-acid RAP1 protein (based on molecular and genetic studies) is shown here. The positions of the DNA-binding domain (Henry et al. 1990) and the putative activation domain (AD) (Hardy et al. 1992a), telomere/silencing domain (TSD), and carboxy-terminal tail domain (TTD) are indicated by the boxes. Also shown are the positions of *rap1* mutations identified to date, including *rap1*[ts] (temperature-sensitive for viability; Lustig et al. 1990; Kurtz and Shore 1991), *rap1*[s] (defective in *HMR* silencing; Sussel and Shore 1991), *rap1*[t] mutations (Kyrion et al. 1992), as well as two nonsense mutations that truncate the terminal 25–28 amino acids (*rap1-20* and *rap1-21*). RIF1, designated by the arrow, interacts with the residues defined by the *rap1*[s] alleles (Hardy et al. 1992b).

the *RIF1* gene result in elongation of telomeres by an average of 250 bp. Association of RIF1 with RAP1 is mediated in part through interactions with amino acids in the region of 725–747 described above. However, since the *rif1* phenotype is less extreme than observed in the *rap1*[t] alleles, it is likely that additional factors associate with RAP1 in this process.

Other telomere-binding proteins in yeast Several additional telomere-binding proteins have been identified in yeast. TBF-α identified by Liu and Tye (1991) binds in vitro to duplex poly($G_{1-3}T$) tracts, although it does not exhibit a specific footprinting pattern. In contrast, TBF-β binds in vitro to both *Tetrahymena* and human telomeric repeats (Liu and Tye 1991). As noted above, human telomeric repeats are present within some XY' junctions and within the Y' element. The gene encoding this protein (*TBF1*) is essential for viability (Brigati et al. 1993). However, *tbf1* mutants do not exhibit any defects in telomere length. An additional protein binding to telomere-like G-rich single-stranded DNA has been identified in yeast. This protein also displays an endonucleoltyic activity that cleaves single-stranded DNA several bases 5' to the G-rich sequence (Liu et al. 1993). The function of this endonuclease is unknown, but it has been proposed as a mechanism for eliminating G-G folded structures prior to telomere addition.

Runge and Zakian (1989) have obtained in vivo evidence for an additional telomere-binding activity by testing the effect of introducing a high copy of telomeric repeats into wild-type strains. They found that telomere tract size increases by ~50% in strains containing 200 copies of 350-bp telomeric repeats. This effect is not end-specific, since it is observed both when tracts are present in circular plasmids and at the termini of linear plasmids. These data suggest that duplex poly($G_{1-3}T$) tracts titrate a factor(s) involved in telomere length control. This factor is unlikely to be RAP1, since loss of RAP1 results in a decrease in telomere tract size.

UNANSWERED QUESTIONS CONCERNING MECHANISMS AFFECTING THE STABILITY OF SIMPLE REPETITIVE DNA SEQUENCES

Our understanding of the mechanisms that affect the stability of simple repeats in yeast is rather primitive. As discussed above, most of the available data indicate that DNA polymerase slippage is likely to be the mechanism that results in instability of poly(GT) tracts, although the possibility of an unusual type of recombination mechanism cannot be excluded. Additional support for a polymerase slippage model might be provided by studies of yeast DNA polymerase slippage during in vitro replication of the poly(GT) tracts. Further studies of mutations affecting the rate of tract alterations are also likely to be informative. It is possible, and perhaps even likely, that tract alterations occur by both mechanisms. For example, it is possible that small changes in tract length occur by DNA polymerase slippage and that large changes occur by recombination. Other questions of interest include: How is the rate of tract instability affected by the number of repeats or the base sequence of repeats? Do tracts that are common in the yeast genome (poly(GT), for example) tend to get larger, and tracts that are uncommon tend to get smaller? Does the chromosomal context influence tract instability? Do alterations in tract size occur during repair DNA synthesis or during replicative DNA synthesis?

The mechanisms involved in replicating telomeric DNA and controlling the length of terminal repeats are even more mysterious. To balance the expected shortening of the telomeres by conventional DNA synthesis, mechanisms that lead to net elongation are required. The only enzymatic mechanism that has been shown in vitro and in vivo to allow net elongation of telomeric sequences is that catalyzed by telomerase (Greider and Blackburn 1985). Numerous searches for this enzyme in yeast have been unsuccessful. Nonetheless, the similar substrate specificity of yeast telomere healing and *Tetrahymena* telomerase remains most consistent with a telomerase-based mechanism (Murray et al. 1988; Harrington and Greider 1991; Lustig 1992; C. Greider, pers. comm.). A second mechanism that (in principle) could allow net elongation of telomeric sequences is gene conversion, a recombination event in which sequences are transferred nonreciprocally from one telomere to another; one such model was proposed by Walmsley et al. (1984). Consistent with this model is the relatively efficient nonreciprocal recombination that occurs between *Tetrahymena* and *Oxytricha* telomeric tracts (poly(G_4T_2) and poly(G_4T_4), respectively) when present at different ends of a linear plasmid (Pluta and Zakian 1989). A third mechanism, which in vitro can lead to net synthesis (Kornberg et al. 1964; Schlotterer and Tautz 1992), is DNA polymerase slippage (Campbell and Newlon 1991). One observation that is supportive of this mechanism is that mutants of DNA polymerase I lead to increased telomere length (Carson and Hartwell 1985); certain mutants of this enzyme also lead to increased rates of putative DNA polymerase slippage events (Ruskin and Fink 1993). In our view, it is likely that the length of telomeric repeats is controlled by a combination of several different mechanisms.

A related question is the involvement of telomere-binding proteins and telomeric chromatin in tract size control. Protein-DNA complexes at the telomere must provide regulated accessibility to the factors that are necessary for maintaining telomere size. Apart from the enzymes involved in the addition process, this may also include recombinational or endonucleolytic mechanisms of telomere loss that act to balance telomere addition. Such complexes may also restrict the accessibility of the telomere to potentially deleterious fusigenic and nucleolytic activities that may lead to chromosome loss and cell death. RAP1 appears to be one component of the telomere that acts in this fashion. It will therefore be interesting to characterize other components that interact with RAP1 in regulating tract size and structure.

Recent studies suggest that the rapid deletion events, first identified in *rap1^t* cells, also occur, albeit more rarely, in wild-type cells that inherit elongated telomeres from heterozygotes (Kyrion et al. 1993). Since the frequency of deletion (via either an endonucleoltyic or recombination pathway) is likely to be dependent on the length of the telomere substrate, rapid deletion events may be an example of a size-dependent regulator of tract length. Further studies are necessary to resolve this issue.

An additional issue concerns interactions between poly(GT) repeats and telomeric repeats. Proof that such interactions occur (at least at low frequency) is the observation that a plasmid-borne poly(GT) tract is altered in length by importing telomeric sequences (Henderson and Petes 1992). One useful approach to examining such interactions is to determine whether mutations that affect one type of repeat also affect the other.

In summary, both internally located and telomeric simple repetitive DNA sequences represent sources of genomic instability in yeast. Studies of poly(GT) tracts indicate that DNA polymerase slippage is likely to be an important mechanism in generating instability. In addition, tract instability is affected by mutations that influence mismatch repair. The control of telomere tract size may be yet more complex, with telomeric sequences being maintained in a dynamic equilibrium between elongation and diminution. This equilibrium appears to be regulated by a number of gene products, including the telomere-binding protein RAP1, DNA polymerase I, and a host of products whose functions remain to be elucidated.

Acknowledgments

We thank S. Henderson and M. Strand for valuable discussions and advice, and E.B. Hoffman for assistance in compiling references. We also thank all those who communicated results prior to publication. These studies were supported by American Cancer Society research grant NP-712 (to T.D.P.), National Science Foundation grant DMB-9120208 (to A.J.L.), and Cancer Center support grant NCI-P30-CA-08748 (to Memorial Sloan-Kettering Cancer Center).

References

Ahn, B.-Y., K.J. Dornfeld, T.J. Fagrelius, and D.M. Livingston. 1988. Effect of limited homology on gene conversion in a *Saccharomyces cerevisiae* plasmid recombination system. *Mol. Cell. Biol.* 8: 2442.

Allshire, R., M. Dempster, and N. Hastie. 1989. Human telomeres contain at least three types of G-rich repeat distributed non-randomly. *Nucleic Acids Res.* 17: 4611.

Allsopp, R., H. Vaziri, C. Patterson, S. Goldstein, E. Younglai, A.B. Futcher, C. Greider, and C. Harley. 1992. Telomere length predicts replicative capacity of human fibroblasts. *Proc. Natl. Acad. Sci.* 89: 10114.

Berg, D.T., J.D. Walls, A.E. Reifel-Miller, and B.W. Grinnell. 1989. E1A-induced enhancer activity of the poly (dG-dT)·(dA-dC) element (GT element) and interactions with a GT-specific enhancer. *Mol. Cell. Biol.* 9: 5248.

Bernards, A., P. Michels, C. Linche, and P. Borst. 1983. Growth of chromosome ends in multiplying trypanosomes. *Nature* 303: 592.

Biessmann, H., S. Carter, and J. Mason. 1990. Chromosome ends in *Drosophila* without telomeric DNA sequences. *Proc. Natl. Acad. Sci.* 87: 1758.

Biessmann, H., L. Champion, M. O'Hair, K. Ikenaga, B. Kasravi, and J. Mason. 1992. Frequent transpositions of *Drosophila melanogaster* HeT-A transposable elements to receding chromosome ends. *EMBO J.* 11: 4459.

Boeke, J.D., F. Lacroute, and G. Fink. 1984. A positive selection for mutants lacking orotidine-5'-phosphate decarboxylase activity in yeast: 5-fluoroorotic acid resistance. *Mol. Gen. Genet.* 197: 345.

Brigati, C., S. Kurtz, D. Balderes, G. Vidali, and D. Shore. 1993. An essential yeast gene encoding a TTAGGG repeat-binding protein. *Mol. Cell. Biol.* 13: 1306.

Buchmann, A., A. Lue, and R. Kornberg. 1988. Connections between transcriptional activators, silencers, and telomeres as revealed by functional analysis of a yeast DNA-binding protein. *Mol. Cell. Biol.* **8:** 5086.

Budarf, M. and E. Blackburn. 1986. Chromatin structure of the telomeric region and the 3'-non-transcribed spacer of *Tetrahymena* ribosomal RNA genes. *J. Biol. Chem.* **261:** 363.

Campbell, J.L. and C.S. Newlon. 1991. Chromosomal DNA replication. In *The molecular biology of the yeast* Saccharomyces: *Genome dynamics, protein synthesis, and energetics* (ed. J.R. Broach et al.), p. 41. Cold Spring Harbor Laboratory Press, Cold Spring Harbor, New York.

Cardenas, M., A. Bianchi, and T. deLange. 1993. A *Xenopus* egg factor with DNA-binding properties characteristic of terminus-specific telomeric proteins. *Genes Dev.* **7:** 883.

Carson, M.J. and L.H. Hartwell. 1985. *CDC17*: An essential gene that prevents telomere elongation in yeast. *Cell* **42:** 249.

Caskey, C.T., A. Pizzuti, Y.-H. Fu, R.G. Fenwick, and D.L. Nelson. 1992. Triplet repeat mutations in human disease. *Science* **256:** 784.

Chan. C. and B.-K. Tye. 1983. Organization of DNA sequences and replication origins at yeast telomeres. *Cell* **33:** 563.

Conrad, M., J. Wright, A. Wolf, and V. Zakian. 1990. RAP1 protein interacts with yeast telomeres *in vivo*: Overproduction alters telomere structure and decreases chromosome stability. *Cell* **63:** 739.

Counter, C., A. Avilion, C. LeFeuvre, N. Stewart, C. Greider, C. Harley, and S. Bacchetti. 1992. Telomere shortening associated with chromosome instability is arrested in immortal cells which express telomerase activity. *EMBO J.* **11:** 1921.

Cross, S., R. Allshire, S. McKay, N. McGill, and H. Cooke. 1989. Cloning of human telomeres by complementation in yeast. *Nature* **338:** 771.

deLange, T., L. Shiue, R. Myers, D. Cox, S. Naylor, A. Killery, and H. Varmus. 1990. Structure and variability of human chromosome ends. *Mol. Cell. Biol.* **10:** 518.

Ferguson, B. and W. Fangman. 1992. A position effect on the time of replication origin activation in yeast. *Cell* **68:** 333.

Ferguson, B., B. Brewer, A. Reynolds, and W. Fangman. 1991. A yeast origin of replication is activated late in S phase. *Cell* **65:** 507.

Fogel, S., J.W. Welch, and E.J. Louis. 1984. Meiotic gene conversion mediates gene amplification in yeast. *Cold Spring Harbor Symp. Quant. Biol.* **49:** 55.

Freund, A., M. Bichara, and R. P. Fuchs. 1989. Z-DNA-forming sequences are spontaneous deletion hotspots. *Proc. Natl. Acad. Sci.* **86:** 7465.

Gottschling, D. and T. Cech. 1984. Chromatin structure of the molecular ends of *Oxytricha* macronuclear DNA: Phased nucleosomes and a telomeric complex. *Cell* **38:** 501.

Gottschling, D. and V. Zakian. 1986. Telomere proteins: Specific recognition and protection of the natural termini of *Oxytricha* macronuclear DNA. *Cell* **47:** 195.

Greider, C.W. 1991. Telomerase is processive. *Mol. Cell. Biol.* **11:** 4572.

Greider, C.W. and E.H. Blackburn. 1985. Identification of a specific telomere terminal transferase activity in *Tetrahymena* extracts. *Cell* **43:** 405.

Hamada, H., M. G. Petrino, and T. Kakunaga. 1982. A novel repeated element with Z-DNA forming potential is widely found in evolutionarily diverse genomes. *Proc. Natl. Acad. Sci.* **79**: 6465.

Hardy, C., D. Balderes, and D. Shore. 1992a. Dissection of a carboxy-terminal region of the yeast regulatory protein RAP1 with effects on both transcriptional activation and silencing. *Mol. Cell. Biol.* **12**: 1209.

Hardy, C., L. Sussel, and D. Shore. 1992b. A RAP1-interacting protein involved in transcriptional silencing and telomere length regulation. *Genes Dev.* **6**: 801.

Harley, C., A.B. Futcher, and C. Greider. 1990. Telomeres shorten during ageing in fibroblasts. *Nature* **345**: 458.

Harrington, L. and C. Greider. 1991. Telomerase primer specificity and chromosome healing. *Nature* **353**: 451.

Hastie, N., M. Dempster, M. Dunlop, A. Thompson, D. Green, and R. Allshire. 1990. Telomere reduction in human colorectal carcinoma and with ageing. *Nature* **346**: 866.

Henderson, E. and E. Blackburn. 1989. An overhang 3' terminus is a conserved feature of telomeres. *Mol. Cell. Biol.* **9**: 345.

Henderson, E., C. Hardin, C. Walk, I. Tinoco, and E. Blackburn. 1987. Telomeric DNA oligonucleotides form novel intramolecular structures containing guanine-guanine base pairs. *Cell* **51**: 899.

Henderson, S.T. and T.D. Petes. 1992. Instability of simple sequence DNA in *Saccharomyces cerevisiae*. *Mol. Cell. Biol.* **12**: 2749.

———. 1993. Instability of a plasmid-borne palindromic repeat in *Saccharomyces cerevisiae*. *Genetics* **134**: 57.

Henry, Y., A. Chambers, J. Tsang, A. Kingsman, and S. Kingsman. 1990. Characterization of the DNA binding domain of the yeast RAP1 protein. *Nucleic Acids Res.* **18**: 2617.

Horowitz, H. and J. Haber. 1984. Subtelomeric regions of yeast chromosomes contain a 36 base-pair tandemly repeated sequence. *Nucleic Acids Res.* **12**: 7105.

Huntington's Disease Collaborative Research Group. 1993. A novel gene containing a trinucleotide repeat that is expanded and unstable on Huntington's disease chromosomes. *Cell* **72**: 971.

Jeffreys, A.J., A. MacLeod, K. Tamaki, D.L. Neil, and D.G. Monckton. 1991. Minisatellite repeat coding as a digital approach to DNA typing. *Nature* **354**: 204.

Johnson, R.E., S.T. Henderson, T.D. Petes, S. Prakash, M. Bankmann, and L. Prakash. 1992. *Saccharomyces cerevisiae RAD5*-encoded DNA repair protein contains DNA helicase and zinc-binding sequence motifs and affects the stability of simple repetitive sequences in the genome. *Mol. Cell. Biol.* **12**: 3807.

Kipling, D. and H. Cooke. 1990. Hypervariable ultra-long telomeres in mice. *Nature* **347**: 400.

Klein, F., T. Laroche, M. Cardenas, J. Hoffmann, D. Schweizer, and S. Gasser. 1992. Localization of RAP1 and topoisomerase II in nuclei and meiotic chromosomes of yeast. *J. Cell. Biol.* **117**: 935.

Klobutcher, L., M. Swanton, P. Donini, and D. Prescott. 1981. All gene-sized DNA molecules in four species of hypotrichs have the same terminal sequence and an unusual 3' terminus. *Proc. Natl. Acad. Sci.* **78**: 3015.

Kornberg, A., L.L. Bertsch, J.F. Jackson, and H.G. Khorana. 1964. Enzymatic synthesis of deoxyribonucleic acid, XVI. Oligonucleotides as templates and

the mechanism of their repication. *Proc. Natl. Acad. Sci.* **51**: 315.
Kramer, W., B. Kramer, M.S. Williamson, and S. Fogel. 1989. Cloning and nucleotide sequence of DNA mismatch repair gene *PMS1* from *Saccharomyces cerevisiae*: Homology of *PMS1* to procaryotic *MutL* and *HexB*. *J. Bacteriol.* **171**: 5339.
Kunes, S., D. Botstein, and M. Fox. 1990. Synapsis-mediated fusion of free DNA ends forms inverted dimer plasmids in yeast. *Genetics* **124**: 67.
Kurtz, S. and D. Shore. 1991. RAP1 protein activates and silences transcription of mating-type genes in yeast. *Genes Dev.* **5**: 616.
Kunkel, T.A. 1986. Frameshift mutagenesis by eucaryotic DNA polymerases *in vitro*. *J. Biol. Chem.* **261**: 13581.
Kyrion, G., K. Boakye, and A. Lustig. 1992. C-terminal truncation of RAP1 results in the deregulation of telomere size, stability and function in *Saccharomyces cerevisiae*. *Mol. Cell. Biol.* **12**: 5159.
Kyrion, G., K. Liu, C. Liu, and A. Lustig. 1993. RAP1 and telomere structure regulate telomere position effects in *Saccharomyces cerevisiae*. *Genes Dev.* **7**: (in press).
Larson, D., E. Spangler, and E. Blackburn. 1987. Dynamics of telomere length variation in *Tetrahymena thermophila*. *Cell* **50**: 477.
Levinson, G. and G. A. Gutman. 1987. High frequency of short frameshifts in poly-CA/GT tandem repeats borne by bacteriophage M13 in *Escherichia coli* K-12. *Nucleic Acids Res.* **15**: 5323.
Liu, Z. and B.-K. Tye. 1991. A yeast protein that binds to vertebrate telomeres and conserved telomeric junctions in yeast. *Genes Dev.* **5**: 49.
Liu, Z., J. Frantz, W. Gilbert, and B.-K. Tye. 1993. Identification and characterization of a nuclease activity specific for G4 tetrastranded DNA. *Proc. Natl. Acad. Sci.* **90**: 3162.
Lohe, A.R. and D.L. Brutlag. 1987. Adjacent satellite DNA segments in *Drosophila*. *J. Mol. Biol.* **194**: 171.
Louis, E. and J. Haber, 1991. Evolutionarily recent transfer of a group I mitochondrial intron to telomere regions in *Saccharomyces cerevisiae*. *Curr. Genetics* **20**: 411.
———. 1992. The structure and evolution of subtelomeric Y' repeats in *Saccharomyces cerevisiae*. *Genetics* **131**: 559.
Lowenhaupt, K., A. Rich, and M.L. Pardue. 1989. Nonrandom distribution of long mono- and dinucleotide repeats in *Drosophila* chromosomes: Correlations with dosage compensation, heterochromatin, and recombination. *Mol. Cell. Biol.* **9**: 1173.
Lundblad, V. and J. Szostak. 1989. A mutant with a defect in telomere elongation leads to senescence in yeast. *Cell* **57**: 633.
Lustig, A. 1992. Hoogsteen G-G base pairing is dispensable for telomere healing in yeast. *Nucleic Acids Res.* **20**: 3021.
Lustig, A. and T.D. Petes. 1984. Long poly(A) tracts in the human genome are associated with the *Alu* family of repeated elements. *J. Mol. Biol.* **180**: 753.
———. 1986. Identification of yeast mutants with altered telomere structure. *Proc. Natl. Acad. Sci.* **83**: 1398.
Lustig, A., S. Kurtz, and D. Shore. 1990. Involvement of the silencer and UAS binding protein RAP1 in regulation of telomere length. *Science* **250**: 549.
McClintock, B. 1941. The stability of broken ends of chromosomes of *Zea mays*. *Genetics* **26**: 234.

McEachern, M. and J. Hicks. 1993. Unusually large telomeric repeats in the yeast *Candida albicans*. *Mol. Cell. Biol.* **13:** 551.

Morin, G. 1989. The human telomere terminal transferase enzyme is a ribonucleoprotein that synthesizes TTAGGG repeats. *Cell* **59:** 521.

Muller, F., C. Wicky, A. Spicher, and H. Tobler. 1991. New telomere formation after developmentally regulated chromosomal breakage during the process of chromatin diminution in *Ascaris lumbricoides*. *Cell* **67:** 815.

Murray, A., T. Claus, and J. Szostak. 1988. Characterization of two telomeric DNA processing reactions in *Saccharomyces cerevisiae*. *Mol. Cell. Biol.* **8:** 4642.

Naylor, L.H. and E.M. Clark. 1990. $d(TG)_n \cdot d(CA)_n$ sequences upstream of the rat prolactin gene form Z-DNA and inhibit gene transcription. *Nucleic Acids Res.* **18:** 1595.

Okazaki, S., K. Tsuchida, H. Maekawa, H. Ishikawa, and H. Fujiwara. 1993. Identification of a pentanucleotide telomeric sequence, $(TTAGG)_n$, in the silkworm *Bombyx mori* and in other insects. *Mol. Cell. Biol.* **13:** 1424.

Olovnikov, A. 1971. Principles of marginotomy in template synthesis of polynucleotides. *Dokl. Akad. Nauk.* **201:** 1496.

Panyutin, I., O. Kovalsky, E. Budowsky, R. Dickerson, M. Rickhirev, and A. Lipanov. 1990. G-DNA: A twice-folded DNA structure adopted by single-stranded oligo(dG) and its implications for telomeres. *Proc. Natl. Acad. Sci.* **87:** 867.

Petes, T.D. and C. Hill. 1988. Recombination between repeated genes in microorganisms. *Annu. Rev. Genet.* **22:** 147.

Petes, T.D., R.E. Malone, and L.S. Symington. 1991. Recombination in yeast. In *The molecular biology of the yeast* Saccharomyces: *Genome dynamics, protein synthesis, and energetics* (ed. J.R. Broach et al.), p. 407. Cold Spring Harbor Laboratory Press, Cold Spring Harbor, New York.

Pluta, A. and V. Zakian. 1989. Recombination occurs during telomere formation in yeast. *Nature* **337:** 429.

Price, C. 1990. Telomere structure in *Euplotes crassus*: Characterization of DNA-protein interactions and isolation of a telomere-binding protein. *Mol. Cell. Biol.* **10:** 3421.

Ripley, L.S. 1990. Frameshift mutation: Determinants of specificity. *Annu. Rev. Genet.* **24:** 189.

Runge, K. and V. Zakian. 1989. Introduction of extra telomeric DNA sequences into *Saccharomyces cerevisiae* results in telomere elongation. *Mol. Cell. Biol.* **9:** 1488.

Ruskin, B. and G.R. Fink. 1993. Mutations in *POL1* increase the mitotic instability of inverted repeats in *Saccharomyces cerevisiae*. *Genetics* **134:** 43.

Schlotterer, C. and D. Tautz. 1992. Slippage synthesis of simple sequence DNA. *Nucleic Acids Res.* **20:** 211.

Sen, D. and W. Gilbert. 1990. A sodium-potassium switch in the formation of four-stranded G4-DNA. *Nature* **344:** 410.

Shampay, J. and E. Blackburn. 1988. Generation of telomere-length heterogeneity in *Saccharomyces cerevisiae*. *Proc. Natl. Acad. Sci.* **85:** 534.

Shampay, J., J.W. Szostak, and E.H. Blackburn. 1984. DNA sequences of telomeres maintained in yeast. *Nature* **310:** 154.

Shippen-Lentz, D. and E. Blackburn. 1990. Functional evidence for an RNA template in telomerase. *Science* **247:** 5462.

Shore, D. and K. Nasmyth. 1987. Purification and cloning of a DNA binding protein from yeast that binds to both silencer and activator elements. *Cell* **51:** 721.
Streisinger, G., Y. Okada, J. Emrich, J. Newton, A. Tsugita, E. Terzaghi, and M. Inouye. 1966. Frameshift mutations and the genetic code. *Cold Spring Harbor Symp. Quant. Biol.* **31:** 77.
Stringer, J.R. 1985. Recombination between poly [d(GT)·d(CA)] sequences in simian virus 40-infected cultured cells. *Mol. Cell. Biol.* **5:** 1247.
Sundquist, W. and A. Klug. 1989. Telomeric DNA dimerizes by formation of guanine tetrads between hairpin loops. *Nature* **342:** 825.
Sussel, L. and D. Shore. 1991. Separation of transcriptional activation and silencing functions of the *RAP1*-encoded repressor/activator protein 1: Isolation of viable mutants affecting both silencing and telomere length. *Proc. Natl. Acad. Sci.* **88:** 7749.
Szostak, J. and E. Blackburn. 1982. Cloning yeast telomeres on linear plasmid vectors. *Cell* **29:** 245.
Tautz, D., M. Trick, and G.A. Dover. 1986. Cryptic simplicity in DNA is a major source of genetic variation. *Nature* **322:** 652.
Teschke, C., G. Sollender, and K. Moritz. 1991. The highly variable pentameric repeats of the AT-rich germline limited DNA in *Parascaris univalens* are the telomeric repeats of somatic chromosomes. *Nucleic Acids Res.* **19:** 2677.
Timmons, M.S., M. Lieb, and R.C. Deonier. 1986. Recombination between IS5 elements: Requirement for homology and recombination functions. *Genetics* **113:** 797.
Treco, D. and N. Arnheim. 1986. The evolutionarily conserved repetitive sequence d(GT·AC)$_n$ promotes reciprocal exchange and generates unusual recombination tetrads during yeast meiosis. *Mol. Cell. Biol.* **6:** 3934.
Tu, H. and M.J. Casadaban. 1990. The upstream activating sequence for L-leucine gene regulation in *Saccharomyces cerevisiae*. *Nucleic Acids Res.* **18:** 3923.
Vignais, M.-L., J. Huet, J.-M. Buhler, and A. Sentenac. 1990. Contacts between the factor TUF and RPG sequences. *J. Biol. Chem.* **265:** 14669.
Wahls, W.P., L.J. Wallace, and P.D. Moore. 1990. The Z-DNA motif (d(TG)$_{30}$) promotes reception of information during gene conversion events while stimulating homologous recombination in human cells in culture. *Mol. Cell. Biol.* **10:** 785.
Walmsley, R. and T.D. Petes. 1985. Genetic control of chromosome length in yeast. *Proc. Natl. Acad. Sci.* **82:** 506.
Walmsley, R.M., J.W. Szostak, and T.D. Petes. 1983. Is there left-handed DNA at the ends of yeast chromosomes? *Nature* **302:** 84.
Walmsley, R.M., C.S.M Chan, B.-K. Tye, and T.D. Petes. 1984. Unusual DNA sequences associated with the ends of yeast chromosomes. *Nature* **310:** 157.
Wang, S.-S. and V. Zakian. 1990. Sequencing of *Saccharomyces* telomeres cloned using T4 DNA polymerase reveals two domains. *Mol. Cell. Biol.* **10:** 4415.
Watson, J. 1972. Origin of concatemeric T7 DNA. *Nature New Biol.* **239:** 197.
Weber, B., C. Collins, C. Robbins, R. Magenis, A. Delaney, J. Gray, and M. Hayden. 1990. Characterization and organization of DNA sequences adjacent to the human telomere associated repeat (TTAGGG)$_n$. *Nucleic Acids Res.* **18:** 3353.

Weber, J.L. and P.E. May. 1989. Abundant class of human DNA polymorphisms which can be typed using the polymerase chain reaction. *Am. J. Hum. Genet.* **44:** 388.

Weiss, R.B., D.M. Dunn, J.F. Atkins, and R.F. Gesteland. 1987. Slippery runs, shifty stops, backward steps, and forward hops: -2, -1, +1, +2, +5, and +6 ribosomal frameshifting. *Cold Spring Harbor Symp. Quant. Biol.* **52:** 687.

Wellinger, R., A. Wolf, and V. Zakian. 1993. *Saccharomyces* telomeres acquire single-strand TG_{1-3} tails late in S phase. *Cell* **72:** 51.

Wells, R., G. Germino, S. Krishna, V. Buckle, and S. Reeders. 1990. Telomere-related sequences at interstitial sites in the human genome. *Genomics* **8:** 699.

Williamson, J., M. Raghuraman, and T. Cech. 1989. Monovalent cation-induced structure of telomeric DNA: The G-quartet model. *Cell* **59:** 871.

Williamson, M.S., J.C. Game, and S. Fogel. 1985. Meiotic gene conversion mutants in *Saccharomyces cerevisiae*. I. Isolation and characterization of *pms1-1* and *pms1-2*. *Genetics* **110:** 609.

Wright, J., D. Gottschling, and V. Zakian. 1992. *Saccharomyces* telomeres assume a non-nucleosomal chromatin structure. *Genes Dev.* **6:** 197.

Yu, G.-L., J. Bradley, L. Attardi, and E. Blackburn. 1990. In vivo alteration of telomere sequences and senescence caused by mutated *Tetrahymena* telomerase RNAs. *Nature* **344:** 126.

Zakian, V. 1989. Structure and function of telomeres. *Annu. Rev. Genet.* **23:** 579.

Defined Ordered Sequence DNA, DNA Structure, and DNA-directed Mutation

Robert D. Wells and Richard R. Sinden

Center for Genome Research, Institute of Biosciences and Technology
Texas A&M University, Texas Medical Center
Houston, Texas 77030-3303

Numerous regions of *defined ordered sequence* DNA (dosDNA) exist in the genomes of prokaryotes and eukaryotes. In contrast to the bulk of the genome, ordered sequences contain bases organized with specific symmetry elements. dosDNA includes inverted repeats, mirror repeats, direct repeats, and mono-, di-, tri-, tetra-, and higher-order nucleotide repeats. dosDNA gives rise to a number of alternate helical forms of DNA including bends, cruciforms, intramolecular triplex structures, Z-DNA, and tetraplex DNA. In some cases, dosDNA plays critical roles in biology. dosDNAs also participate in spontaneous, specific DNA-directed mutagenic events that can lead to cancer and human genetic diseases. The expansion of triplet repeats leading to the fragile X syndrome, myotonic dystrophy, Kennedy's disease (spino-bulbar muscular atrophy), and Huntington's disease may involve aberrant events occurring during DNA replication.

Main points discussed include:

❑ dosDNA and alternate DNA conformations

❑ dosDNA and DNA-directed mutation

❑ human genetic disorders involving the expansion of triplet repeats

❑ mutagenesis of triplet repeats: possible mechanisms

INTRODUCTION

During the last 10–15 years, our understanding of DNA as a structurally heterogeneous molecule has grown considerably. The canonical Watson-Crick right-handed double helix represents the predominant form of DNA. However, under physiological conditions of supercoiling and divalent cations, a large number of alternate helical forms of DNA can exist within regions of nonrandom or ordered DNA sequence. Most of these alternate helical forms of DNA are important biologically. Regions of DNA containing stable curvature have been shown to be involved in various biological reactions of DNA, including replication, site-specific recombination, and gene regulation (Hagerman 1990). Cruciforms, which can form within inverted repeats at high levels in vivo (Zheng et al. 1991), may be involved at origins of replication (Noirot et al. 1990; Ward et al. 1990). Cruciforms can also promote deletion between flanking direct repeats (Collins 1980; Dasgupta et al. 1987; Pierce et al. 1991; Sinden et al. 1991; Weston-Hafer and Berg 1991). Z-DNA and intramolecular triplex structures can exist in vivo (Jaworski et al. 1987; Zheng et al. 1991; Rahmouni and Wells 1992; Ussery and Sinden 1993) and may influence genetic recombination and the regulation of gene expression. Simple direct repeats, which can form a number of different alternate DNA conformations, have been shown to be genetically unstable in the human genome (Weber 1990; Morral et al. 1991; Edwards et al. 1992). Massive expansion in the length of triplet repeats has been shown to be the etiology of several human genetic diseases (Caskey et al. 1992; MacDonald et al. 1993).

The accurate replication of the human genome is critical for procreation. dosDNA presents special problems for the replication machinery. Deletion, duplications, correction of quasi-palindromes, and rearrangements are frequently associated with sites of directed ordered DNA sequence (Krawczak and Cooper 1991). Misalignment during DNA replication can lead to deletions and duplications of DNA. Direct repeats, especially runs of di- or trinucleotide repeats, lead to frequent slippage during replication, promoting expansion or contraction of the repetitive element. The formation of hairpin stems, intramolecular triplexes, or other non-B-DNA structures may present blocks to the replication fork, leading to an increased probability of misalignment-based mutagenesis or genetic recombination at these ordered sequences. Sequences with palindromic or quasi-palindromic symmetry can also promote intermolecular strand switching in which a nascent progeny strand switches from one to the other template strand.

Recently, the massive expansion of triplet repeats has been shown to occur with non-Mendelian genetics, leading to human genetic diseases that show a phemonenon known as *anticipation*, in which the severity of the disease increases with each successive generation (Caskey et al. 1992; Richards and Sutherland 1992; Sinden and Wells 1992). The expansion of triplet repeats represents a novel type of mutational event. In some cases, the size of the repeat increases by a factor of 10 or more as the mutation is passed from parent to child. The molecular etiology of this type of mutational event is unknown (Sinden and Wells 1992). In this paper, we consider possible mechanisms of expansion based on our current understanding of the alternate conformations of DNA adopted by, and molecular mechanisms of mutation associated with, dosDNAs.

dosDNA AND ALTERNATE DNA CONFORMATIONS

A wide variety of investigations have been conducted on non-B-DNA structures, especially during the past 15 years, that have revealed a range of interesting conformational features. Excellent reviews have been written on some of these conformations, and the reader is referred to them for further details. This chapter contains general information that may aid in the comprehension of the potential role of these structures in mutagenesis and replication as related to human diseases.

In general, all of these conformational features represent underwound DNA primary helices and are stabilized by negative supercoiling. The only exceptions are tetraplex DNA and bent (curved) DNA, which are described in linear fragments.

Left-handed Z-DNA

Left-handed Z-DNA (Fig. 1) is the most widely investigated non-B-DNA structure. It is formed within regions of alternating purine and pyrimidine sequences such as (GT) or (GC) under physiological conditions in negatively supercoiled DNA (Wells 1988; Rich et al. 1984). A wide range of investigations have been conducted on Z-DNA, including the following: the conditions that stabilize the structure; the effect of methylation of cytosine residues; types of sequences that adopt Z helices and the rules governing their formation; the properties of B-Z junctions; the development of new assays for Z structures; features of the immunological properties of different left-handed structures; the characteristics of proteins that bind to Z helices; thermodynamics and kinetics of transitions; and interactions with small ligands such as carcinogens and mutagens. Left-handed Z-DNA exists in living cells (Jaworski et al. 1987) and is stabilized by domains of negative supercoiling (Zheng et al. 1991; Rahmouni and Wells 1992).

Cruciforms and slipped structures

Cruciforms exist at segments of inverted repeat sequences in negatively supercoiled DNA (Fig. 1) (Wells 1988). These conformations have been

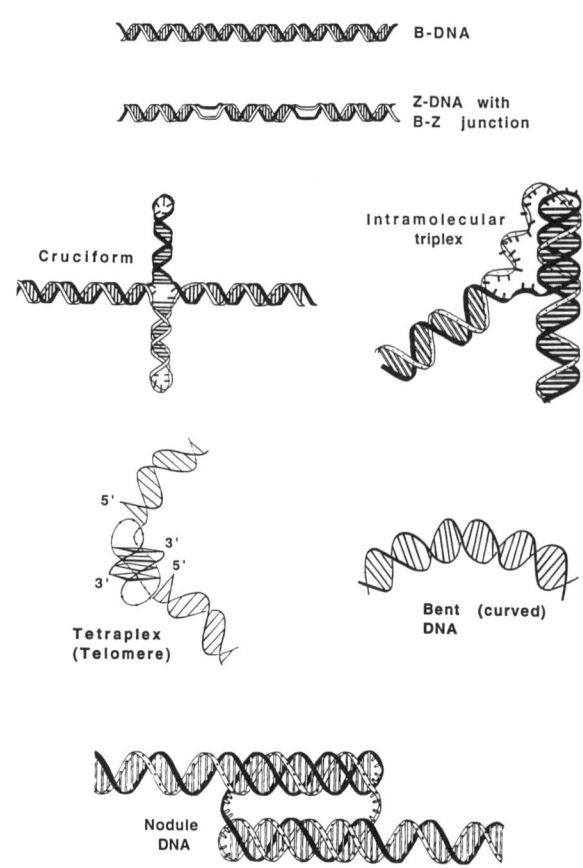

Figure 1 Models of the structures of Z-DNA with B-Z junctions, a cruciform, an intramolecular triplex, a tetraplex, bent DNA, and nodule DNA are shown. Descriptions of these DNA structures and pertinent references are included in the text.

widely investigated as models for Holliday structures related to genetic recombination (Murchie et al. 1992).

Cruciforms can exist at high levels in bacterial cells, but their level of existence is dependent on their intra- or extragenic location and the physiological conditions of the cell, as well as the level of transcription or replication of the DNA (McClellan et al. 1990; Zheng et al. 1991). Substantial investigations are described elsewhere in this chapter on the instability of inverted repeats and mutations that occur related to this conformational feature (for review, see Sinden and Wells 1992).

DNA slipped structures have been postulated (McKeon et al. 1984) to occur at direct repeat sequences, in general, upstream of important regulatory sites such as the promoters for mouse α2(I) collagen genes. In principle, all direct repeat sequences can form slipped structures. However, if numerous direct repeat sequences exist, a family of slipped structures can exist with different regions of unpairing. Hence, the characterization of these structures has been difficult. Compared to the other structures described herein, slipped structures have been less thoroughly studied. However, with the discovery of repeating triplet sequences related to several human diseases (fragile X, Huntington's disease, etc.), these conformational features may take on heightened significance.

Triplexes and nodule DNA

Intramolecular DNA triplexes (Fig. 1) occur at tracts of oligopurines in negatively supercoiled DNAs. Mirror repeat sequences are beneficial, and pH values below 7 are stabilizing, since the Hoogsteen configuration with a CGC triplet requires protonation (Wells et al. 1988). Substantial investigations have been conducted on the following features: the effect of sequence and the type of Pur·Pyr sequences; the effect of pH; the influence of methylation of C residues; the types of bi-triplexes that may be formed; the effect of interposing non-Pur·Pyr sequences (Klysik 1992); the influence of environmental factors on the stabilization of the four triplex isomers (Kang and Wells 1992); triplex stabilization by intercalating agents; and related investigations. Interestingly, a DNA conformational alteration (bend) is induced by a neighboring oligopurine tract in the regulatory region of human papillomavirus type 11 (Hartman et al. 1992).

A recent intriguing study by Ulrich et al. (1992) reveals the likely involvement of an intramolecular triplex associated with hereditary persistence of fetal hemoglobin. These investigators describe a secondary DNA structure upstream of the human gamma globulin genes; this structure was found at a homopurine tract and was stabilized by low pH and negative supercoiling in plasmid DNAs. Nuclease sensitivity and binding studies with exogenous oligomers enabled the mapping of the secondary structure with five different natural point mutations. Four of these mutations dramatically reduced the stability of the DNA secondary structure and, therefore, may represent a novel class of genetic defects that alter gene expression by changing the interaction of a critical regulatory molecule with a DNA secondary structure.

Several types of intramolecular bi-triplexes occur depending on their sequences and the environmental conditions (Shimizu et al. 1990). In addition, under appropriate conditions, an interesting new conformation termed nodule DNA (Fig. 1) was described which involves a strand exchanged bi-triplex (Kohwi-Shigematsu and Kohwi 1991; Panyutin and Wells 1992). The formation of this recombination-like DNA structure does not require the presence of proteins but simply the appropriate sequence under the stabilizing conditions.

Tetraplexes

Recently, much attention has been devoted to four-stranded DNA structures (Fig. 1) because of their presence at the telomeric ends of eukaryotic chromosomes. The nucleotide sequence of the telomeric repeat is very similar in organisms as diverse as *Tetrahymena* (T_2G_4), *Oxytricha* and *Stylonychia* (T_4G_4), yeast (TG_{1-3}), plants (T_3AG_3), and humans (T_2AG_3). Substantial studies of stable complexes in solution reveal a four-base interaction between the G residues in oligonucleotides containing short, single runs of three or more Gs. Potassium ions excessively stabilize the fold-back, intermediate structures (Sundquist and Klug 1989; Panyutin et al. 1990; Raghurman and Cech 1990; Sen and Gilbert 1990). The kinetics and thermodynamics of the process have been investigated; the folding of DNA is inhibited by methylation (Raghurman and Cech 1990). The crystal structure of the four-stranded *Oxytricha* telomeric DNA sequence ($G_4T_4G_4$) has been crystallized, and the four-stranded helical structure with Ts in the loops and G residues held together by cyclic hydrogen bonding has been described (Kang et al. 1992).

Anisomorphic DNA at direct repeat sequences

Investigations on plasmids containing other types of direct repeat sequences from herpes simplex virus type-1 (HSV-1) revealed the formation of a novel type of non-B-DNA structure. The site of segment inversion of HSV-1 consists of multiple repeats (from 8 to 23 copies) of a sequence very rich in G+C (92%) with a high bias for purines on one of the strands (92%). Conformational investigations (Wohlrab et al. 1987; Wohlrab and Wells 1989) revealed a family of non-B-DNA conformations called anisomorphic DNA. The presence of this novel conformation was dependent on the number of the DR2 repeats; the 19-mer (228 bp total) and the 14-mer (168 bp) readily formed the alternate structure, whereas pentamer, trimer, and dimer repeats show different properties. S1 and P1 nuclease studies revealed that the new conformation had a major structural aberration at its center and conformational periodicities that were not identical on the complementary strands. In addition, the effect of salt and pH, the location of reaction with bromo- and chloracetaldehyde, the type of sequence (direct repeat) involved, and the nature and extent of supercoil-induced relaxations demonstrated that this structure differed from previously recognized conformations, including left-handed Z helices, cruciforms, bent DNA, and slipped structures. We proposed (Wohlrab et al. 1987) the existence of a novel conformation, anisomorphic DNA, with different structures on the complementary strands that elicited a structural aberration at the physical center of the tandem repeats. Since the oligopurine·oligopyrimidine sequence may be inherently inflexible, this supercoil-induced structural change and the physical stress on these inserts in recombinant plasmids tended to bend or crack the DR2 sequences at their centers.

Other investigations showed that plasmids carrying eight contiguous copies of DR2 sequences underwent a series of supercoil-driven conformational transitions, resulting in different extents of relaxation at pH 5.0 (Wohlrab and Wells 1989). These transitions depended on the presence of an appropriate concentration of divalent cations (Mg^{++} and Ca^{++}), which seem to interact specifically with the alternate structure(s). The transitions occurred at approximately the same superhelical density for all lengths of inserts studied. However, the onset of the transition can be shifted to lower negative superhelical densities by increasing NaCl concentrations. This leads to a reduction of the cooperativity of the transition, which takes place over a range of linking isomers under these conditions. Extrapolating from these results, we established physiological conditions where the alternate DNA structure is found at negative superhelical densities as low as –0.035.

This DR2 segment was specifically cleaved by a virus-induced nuclear endonuclease (Wohlrab et al. 1991). Nuclear extracts from several tissue culture cell lines (human, primate, and murine) contain an endonuclease that specifically cleaves sequences at the HSV-1 segment inversion site. Mapping studies identified the preferential site of cleavage as a set of tandemly repeated dodecamers, the DR2 repeats. In addition to the HSV-1 DR2 repeats, certain other G+C-rich sequences with an asymmetric distribution of purines and pyrimidines on the DNA strands, and with appropriate sequences and lengths, were substrates for the nuclease. These data indicated that target site recognition by the enzyme was conformation-specific rather than sequence-specific. Additional recent investigations (Bruckner et al. 1992) are in concert with these interpretations.

Bent (curved) DNA

Bent, or curved, DNA (Fig. 1) was originally discovered at an A-B DNA junction that was distorted by as much as 26° due to the differences in the base-pair planes of the two conformers. Bent DNA occurs with phased sequences where locally curved A tracts must be aligned in phase with the helix repeat in order to produce macroscopic curvature (curvature observed on gels). Thus, runs of A sequences that occur with a periodicity of the helical repeat are particularly likely to form bent conformations. A thorough and recent review on DNA curvature is available (Hagerman 1990) that provides an excellent overview of the relationship between sequence and curvature, static curvature versus anisotropic flexibility, X-ray diffraction and NMR structural studies on A-tract-containing oligomers, the magnitude and direction of the curvature, and the biological relevance of DNA bending.

The most compelling evidence for the involvement of any B-DNA structure in a biological process may be the involvement of bent DNA with the *Escherichia coli* integration host factor (IHF), which is required for efficient integrative recombination of bacteriophage λ. Stenzel et al. (1987) showed that the IHF of *E. coli* binds to bent DNA at the origin of replication of a suitable plasmid. Electrophoretic analyses showed that the IHF bound to bent DNA, and that the protein binding further enhances the degree of bending. More recent investigations by Robertson and Nash (1988) confirmed that the binding of IHF creates bends in DNA in order to help attP condense into a compact structure that is activated for recombination. These workers discovered that IHF binding to either of two sites found within attP did produce bending of DNA. In contrast, the other recombination protein needed for integrative recombination, Int, does not appreciably bend the DNA to which it is bound. These investigators studied protein-directed bends in DNA by electrophoretic mobility of a set of permuted DNA fragments in the presence or absence of IHF and/or Int. More recent studies (Goodman et al. 1992) showed that the IHF protein could be replaced by HU protein or by sequence-directed bends. The strategy of these workers has been to construct chimeric attachment sites in which the IHF-binding sites are replaced by an alternative source of deformation. This work reinforces the interpretation that IHF functions primarily as an architectural element.

dosDNA AND DNA-DIRECTED MUTATION

Ordered DNA sequences present a particular problem to the replication machinery of the cell. For example, repetitive sequence elements promote a high probability of template slippage, which can lead to deletions and duplications as well as strand switching during replication. Inverted repeats in DNA are known to be unstable, giving rise to frequent deletion. Inverted repeats can also promote strand-displacement synthesis. Regions of homopurine·homopyrimidine sequence containing mirror repeat symmetry can act as terminations to replication, and replication pause sites are potentially mutagenic. In this section, we review our current understanding of molecular mechanisms involved in the preferential mutagenesis at ordered DNA sequences.

Polymerase misbehavior: Base substitutions and single base frameshifts

Base substitutions are thought to occur frequently by template miscoding or *misincorporation*. This can occur by an ionization of a base such that it pairs in a non-Watson-Crick fashion, where A does not pair with T and C does not pair with G (Sowers et al. 1986, 1989). In addition, wobble pairing or pairing in an *anti-syn* fashion can lead to transitions or transversions. As shown in Figure 2A, the formation of an A·G$_{syn}$ base

A. Misincorporation (substitution)

5' CTGCTGCTG 3'
3' GACGAA   5'

⇓

5' CTGCTGCTG 3'
3' GACGAAGAC 5'
         *

B. Misincorporation-misalignment (frameshift)

5' CTGCTGCTG 3'
3' GACGACA   5'
          i

⇓
       C
5' CTGCTGTG 3'
3' GACGACAC 5'

C. Misalignment-incorporation (frameshift)

5' CTGCTGCTTG 3'
3' GACGACGA   5'

⇓
        T
5' CTGCTGCTG 3'
3' GACGACGAC 5'

D. Dislocation (substitution)

       T
5' CTGCTGCTG 3'
3' GACGACGAC 5'

⇓

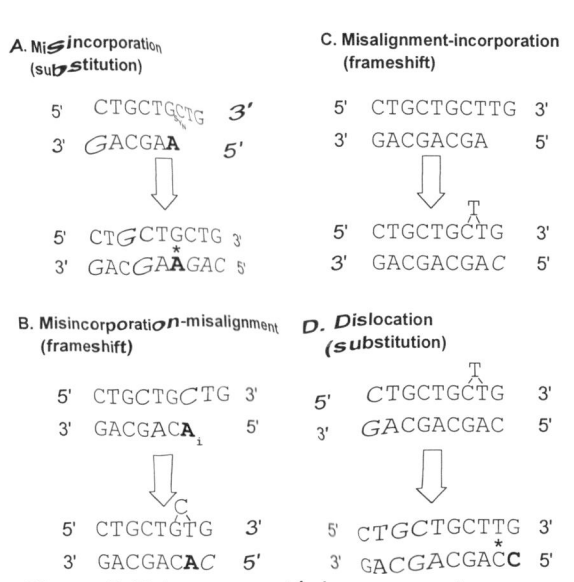

5' CTGCTGCTTG 3'
3' GACGACGACC 5'
          *

Figure 2 Polymerase misbehavior. (The term "polymerase misbehavior" has been used by Thomas A. Kunkel to describe the mutagenic activity of DNA polymerases at the 3' end of a nascent DNA chain.) (A) Base substitutions caused by misincorporation. In the upper example, an A pairs with a G in the syn conformation in the template strand. Extension from this A·G$_{syn}$ base pair (a G·A mispair) results in a base substitution mutation. The G·A mismatch is designated by an asterisk. Bases synthesized after the mutagenic event are in italics. (B) Mechanism by which a frameshift mutation is induced by misincorporation followed by misalignment. In this example, an ionized form of A pairs with the C. In situations where the C is followed on the 3' side by a T, the non-ionized form of A could pair with the T following misalignment. For this misalignment to occur, an extrahelical C would be formed in the template strand. Extension from this A·T pair results in the deletion of a C·G base pair. (C) Introduction of a frameshift mutation by misalignment followed by incorporation. Misalignment occurs with an extrahelical T in the template strand. Continued synthesis results in the deletion of a T·A base pair from the DNA. (D) Dislocation mechanism for base substitution mutations. Dislocation involves the misalignment shown in C. Following the misalignment and incorporation at the misaligned template, the template realigns back to its original register. However, the template now contains a T·C mismatch, and continued polymerization results in a transversion of a T·A base pair to a G·C base pair.

pair will lead to a GC to TA transversion. Misincorporation followed by misalignment can lead to frameshifts, as shown in Figure 2B. Frameshifts can also occur by a misalignment followed by incorporation. In the example shown in Figure 2C, incorporation of an A followed by continued synthesis following misalignment leads to the deletion of an AT base pair. Dislocation, described by Kunkel (1990, 1992), can lead to base substitutions. For dislocation to occur, DNA polymerase dissociates from the 3' end of a nascent progeny strand, and a misalignment occurs in which the 3' end of the nascent progeny strand slips forward or slips backward. Following misalignment, a single base is added to the end of the progeny strand (Fig. 2D). The progeny strand then rehybridizes with the template strand in its original register. When a mismatch exists at the 3' end of the nascent progeny strand, continued synthesis from this mismatch will lead to a base substitution mutation. Extension from mismatched bases occurs with reasonably rapid kinetics (Creighton et al. 1992; Joyce et al. 1992). In most cases, dislocation mutations are evident by the templating of the mutated base from an adjacent or nearby sequence flanking the mutation.

Slipped mispairing

Slipped mispairing or misalignment mutagenesis was suggested in 1966 by Streisinger and colleagues to explain the spontaneous mutagenesis at a run of a single base (Streisinger et al. 1966). Runs of homopurines or homopyrimidines have been identified as hot spots of mutagenesis in a number of bacterial and eukaryotic sources (Streisinger and Owen 1985; Kunkel 1990; Ripley 1991). Moreover, di-, tri-, and tetranucleotide repeats frequently show heterogeneous size distributions in human populations (Weber 1990; Morral et al. 1991; Caskey et al. 1992; Edwards et al. 1992). Mechanisms of mutagenesis involving slipped misalignment during DNA replication likely give rise to these mutagenetic events. Figure 3 shows the molecular mechanisms involved in slipped misalignment. The 3' end of the nascent progeny strand unpairs from the template strand and repairs in a register that is different from the original base-pairing scheme. If the 3' end of the progeny strand slips forward and replication ensues, a heteroduplex will occur with an extra base(s) in the template strand. Upon a second round of replication, synthesis on the progeny strand will lead to duplex DNA containing the deletion. If the 3' end of the progeny strand slips backward before replication ensues, a duplication will result. Slipped misalignment can occur at the level of single nucleotides or several nucleotides. Misalignment can also occur between direct repeats (Fig. 3) that are located hundreds of base pairs apart, resulting in the deletion or duplication of large regions of genetic information (Farabaugh et al. 1978; Levinson and Gutman 1987; Kunkel 1990; Ripley 1991).

Within repeats

(a) Deletion mechanism

```
5' CTGCTGCTG         3'
3' GACGACGACGACGACGAC 5'
```
⇩
```
5' CTGCTGCTG         3'
3' GACGACGACGAC 5'
        G C
         \/
         A
```
⇩
```
5' CTGCTGCTGCTG 3'
3' GACGACGACGAC 5'
      G C
       \/
       A
```

(b) Duplication mechanism

```
5' CTGCTGCTGCTG      3'
3' GACGACGACGACGAC 5'
```
⇩
```
         T
        /\
        C G
5' CTGCTGCTG         3'
3' GACGACGACGACGAC 5'
```
⇩
```
         T
        /\
        C G
5' CTGCTGCTGCTGCTG 3'
3' GACGACGACGACGAC 5'
```

Between direct repeats

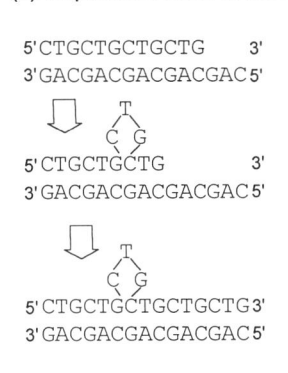

(a) Deletion mechanism

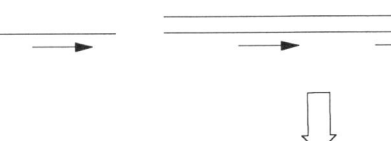

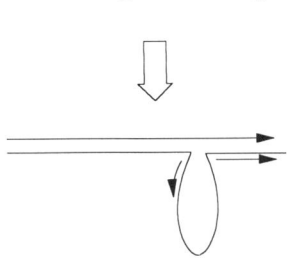

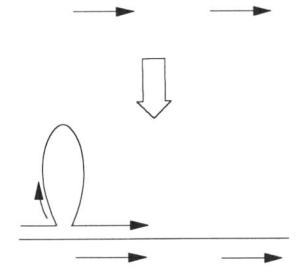

(b) Duplication mechanism

Figure 3 Misalignment-based mutations. *Within repeats:* Misalignment or slipped mispairing within runs of single bases or within triplet repeats as shown above can cause additions or deletions of repeating units. For deletions to occur, misalignment within direct repeats requires a forward slippage of the progeny strand followed by continued synthesis (a). (b) The mechanism for duplication mutations, which involves a backward slippage of the progeny strand followed by continued synthesis. *Between repeats:* Slipped mispairing or misalignment can also cause deletions between direct repeats. In this case, synthesis of the first copy of a direct repeat occurs (direct repeats are denoted by the arrows beneath the bottom DNA strand). This is followed by a forward slippage in the progeny strand pairing the first copy of the direct repeat in the progeny strand with a second copy of the direct repeat in the template strand. Continued synthesis leads to deletion of one copy of the direct repeat and DNA between the direct repeats. (b) Mechanism of duplication between direct repeats. In this case, both copies of the direct repeat and the intervening DNA are synthesized. There is then a backward slippage of the progeny strand resulting in pairing of the second copy of the direct repeat in the progeny strand with the first copy of the direct repeat in the template strand. Continued synthesis leads to a duplication of one copy of the direct repeat and the DNA between the direct repeats. (Adapted, with permission, from Sinden and Wells 1992.)

Slipped mispairing and DNA secondary structure

DNA secondary structures influence deletions between direct repeats
Deletion between direct repeats is frequently associated with additional symmetry elements flanking the direct repeats. Glickman and Ripley suggested that the formation of hairpin structures or other DNA secondary structures flanking the direct repeats may stabilize a misalignment event leading to a specific mutational event (Drake et al. 1983; Glickman and Ripley 1984). Recent tests of this model have confirmed that the formation of hairpin structures forming a perfect three-way junction (see Fig. 4a) will stabilize the misalignment, leading to deletions (Trinh and Sinden 1991, 1993). Moreover, in some cases, the formation of an alternate secondary structure at a replication fork can prevent misalignment and deletion mutagenesis (Trinh and Sinden 1993). In one case, the formation of an imperfect three-way junction apparently does not allow a 3'end of a progeny strand to misalign and undergo a deletion event. Therefore, alternate DNA secondary structures can either increase or decrease the frequency of deletions between direct repeated DNA sequences.

DNA secondary structures influence duplications between direct repeats The presence of dosDNA between direct repeats can have a significant influence on the frequency of particular spontaneous duplication between direct repeats. For a duplication mutation to occur, a large region of DNA between two direct repeats must unwind, and the second copy of the direct repeat in the progeny strand must pair with the first copy of the direct repeat in the template strand. Continued synthesis of the progeny strand will lead to a duplication of genetic information (Fig. 3). Duplication mutations occur readily when there are no dosDNA sequence elements between the direct repeats. However, the presence of inverted repeats flanked by the direct repeats significantly reduces the frequency of duplication mutation (Trinh and Sinden 1993). Hairpin formation in the progeny strand that could facilitate misalignment, leading to a duplication, may occur concomitantly with hairpin formation in the template strand. This will lead to re-pairing of the second copies of the direct repeats in the progeny and template strand. Moreover, the presence of multiple short internal direct repeats has also, in some cases, been shown to reduce the frequency of duplications between flanking direct repeats, compared to a situation in which there is no ordered DNA sequence present between direct repeats (Trinh and Sinden 1993).

Cruciform structures and deletion

Inverted repeats, especially long inverted repeated chromosomal DNA, are unstable in *E. coli*. Cloned inverted repeats in which the inverted

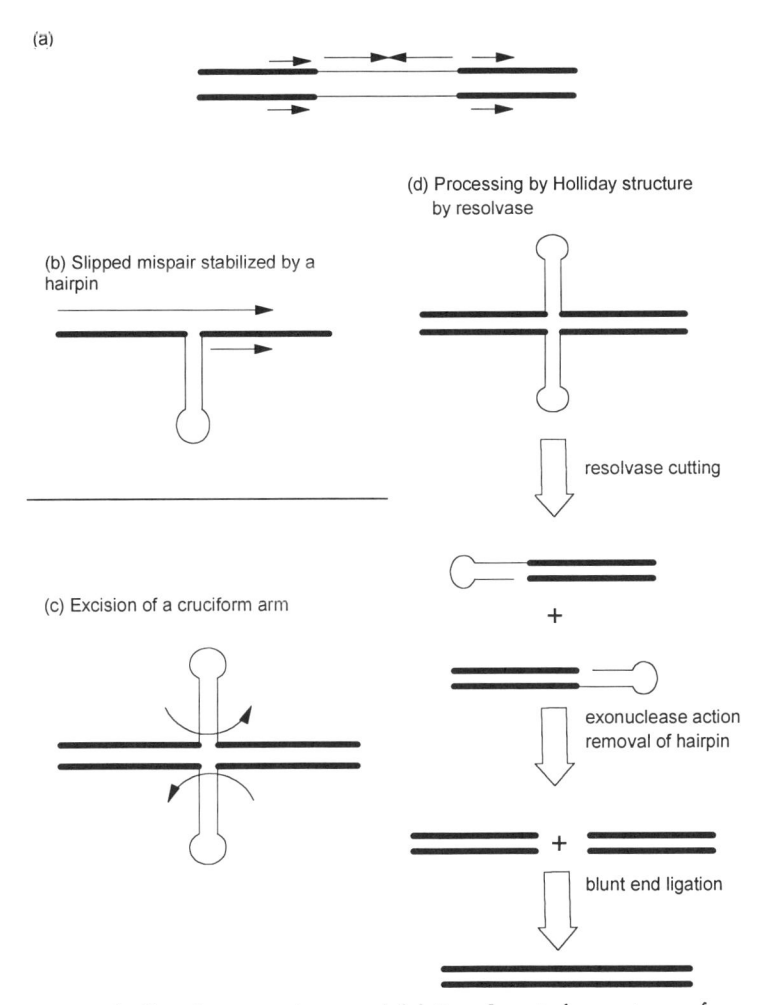

Figure 4 Cruciform structures and deletion. Inverted repeats can form cruciform structures in vitro and in vivo. Long inverted repeats (>100 bp) cloned into bacterial DNA are unstable. Cloned inverted repeats (thin lines in a) are flanked by direct repeats (the restriction sites) which are also part of a larger inverted repeat region. The inverted repeat is denoted by the facing arrows, the direct repeats by the right-facing arrows. The instability of inverted repeats may be the result of several different potential mutagenic mechanisms. (b) For cloned inverted repeats in which an inverted repeat is flanked by direct repeats, a misalignment event can be stabilized by the formation of a hairpin arm in the template strand. (c) Potential excision of cruciform arms. To date, however, no known enzymatic activity will cut off a cruciform arm. (d) Potential mechanism of deletion of inverted repeats involving processing of a Holliday recombination intermediate (a cruciform structure). A number of enzymes are known to introduce cuts at 4-way junctions. This would resolve a cruciform structure into a linear molecule with two hairpin ends. Exonuclease action, removing the hairpins, would result in formation of blunt-end double-strand molecules which, when ligated together, would result in the deletion of the inverted repeat. (Adapted, with permission, from Sinden and Wells 1992.)

repeat is flanked by direct repeats are very unstable (Collins 1980; Ripley 1991; Sinden et al. 1991). It is difficult to maintain cloned inverted repeats greater than about 150 base pairs in *E. coli*. The stability of cloned inverted repeats is a function of length of the direct repeats, length of the hairpin arm, thermal stability of the hairpin stem, and the existence of cruciforms in living cells (Weston-Hafer and Berg 1989; Pierce et al. 1991; Sinden et al. 1991). The formation of a cruciform in living cells presents a substrate for preferential mutagenesis by DNA polymerization. DNA polymerase can stop when it encounters a hairpin stem. Once stopped, there may be a high probability of misalignment between flanking direct repeats. The misalignment may be stabilized by a hairpin structure, leading to a high frequency of deletion (Fig. 4a). Inverted repeats that are not flanked by direct repeats do not exhibit the same instability as inverted repeats flanked by direct repeats. In fact, inverted repeats or very short direct repeats that are not flanked by direct repeats appear to be very stable in the few cases that were examined (Pierce et al. 1991; T.Q. Thinh and R.R. Sinden, unpubl.). Although cruciforms may be unstable because of endonucleolytic attack by Holliday structure resolvases, there is little evidence for the action of resolvases on cruciform structures in *E. coli* cells. This would suggest that there are no mechanisms in *E. coli* for excising the cruciform arm endonucleolytically. In addition, if inverted repeats not flanked by direct repeats are stable genetically, this would suggest that DNA polymerase cannot easily skip over a hairpin arm during replication.

The genetic instability of potential intramolecular triplex-forming regions

Regions of DNA that are known to form intramolecular triplex structures are unstable in *E. coli* when cloned in long stretches (Jaworski et al. 1989; S. Vendetti and R. Wells, unpubl.). There are several reasons that these sequences might be unstable in vivo. First, intramolecular triplex structures have been shown to act in vitro and in vivo as termination sequences for DNA replication (Baran et al. 1991; Brinton et al. 1991). The pausing of a replication fork has been associated with the high frequency of mutagenetic events (Bebenek et al. 1989, 1993; Kunkel 1990; Papanicoleau and Ripley 1991). Therefore, arrest of replication by formation of a triplex structure at the 3' end of a nascent progeny strand may be responsible for the mutations associated with intramolecular triplex-forming regions of DNA. Second, many intramolecular triplex-forming regions, which require polypurine·polypyrimidine sequence motifs, i.e., $(G)_n \cdot (C)_n$ and $(GA)_n \cdot (TC)_n$, have a high probability of simple slipped misalignment, which would give rise to deletions or duplications during DNA replication (see, e.g., Freund et al. 1989). Third, the formation of an alternate secondary structure in vivo could lead to the genetic instability. It has recently been shown that intramolecular triplex DNA structures can exist in living *E. coli* under certain conditions (Ussery and Sinden 1993).

Strand-switch mechanisms of DNA mutagenesis

There are a number of types of mutagenetic events observed in prokaryotes and eukaryotes that may involve strand switching during DNA replication (Ripley 1982, 1991; Krawczak and Cooper 1991). A high frequency of intermolecular strand switching at inverted repeats has been demonstrated in *E. coli* (Ohshima et al. 1992). The correction of a quasi-palindrome to a perfect palindrome may occur by an intermolecular or intramolecular strand switch (Ripley 1982). An intermolecular strand switch involves dissociation of the progeny strand from the template strand and rehybridization with the opposite template strand. For example, a leading strand replication fork may dissociate and cause the 3'end of the progeny strand to rehybridize with the lagging strand template. Intramolecular switch involves dissociation of the progeny strand from the template strand and rehybridization of the 3'end of the progeny strand upon itself, leading to synthesis from the newly synthesized progeny strand, creating a hairpin stem (an inverted repeat). These models are illustrated in Figure 5.

HUMAN GENETIC DISORDERS INVOLVING THE EXPANSION OF TRIPLET REPEATS

Dramatic advances have been made recently on the molecular basis of the genetics of several human diseases. Hereditary unstable DNA provides a new and exciting explanation for some old and perplexing genetic questions. The majority of the chapters in this volume deal with genetic considerations related to these disease states. Hence, we present only a general overview of the subject as it pertains to this chapter.

Fragile X syndrome

Fragile X syndrome is the most frequent inherited mental retardation in humans and segregates as an X-linked dominant disorder with reduced penetrance, since either sex, when carrying the fragile X mutation, may exhibit mental deficiency (for review, see Sutherland et al. 1991; Verkerk et al. 1991; Caskey et al. 1992; Harley et al. 1992; Rennie 1992; Richards and Sutherland 1992.) This human disease syndrome with mental retardation and minor dysmorphic features is so-called because most affected individuals exhibit cytogenetically a fragile site at the locus of this disease on the X chromosome. The mutation responsible for fragile X syndrome was first characterized as a heritable unstable DNA sequence

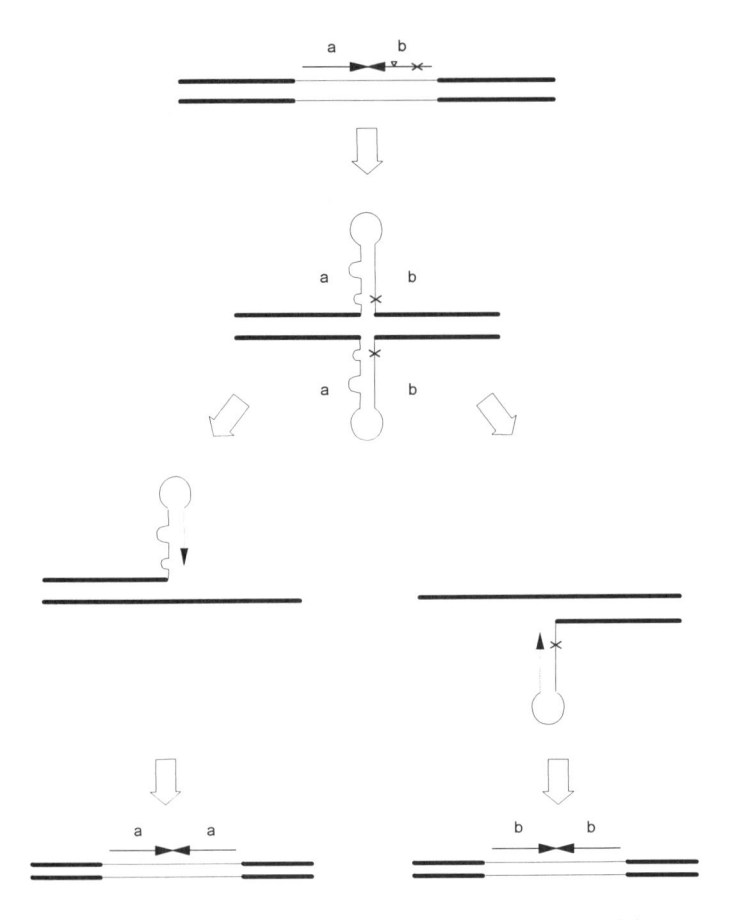

Corrects to aa **Corrects to bb**

Figure 5 Correction of quasi-palindrome mutation. The quasi-palindrome is an imperfect inverted repeat. This is illustrated by the facing arrows of two sides, a and b. The triangle in b represents a small deletion relative to a, whereas the X represents a single base change relative to arm a. If a cruciform formed, there would be a mismatch at the single base change X and a loop-out structure at the position of the deletion in arm b. Quasi-palindromes become corrected to perfect palindromes in a single mutagenic event. This can occur by two mechanisms. One involves an intramolecular strand switch. On the left is shown an intramolecular strand switch in which replication occurs down the hairpin arm. Arm a is templated, resulting in the correction to an aa inverted repeat. Correction using the other strand of the sequence would result in a bb inverted repeat. The correction of quasi-palindrome can occur by a second mechanism that involves an intermolecular strand switch in which polymerase switches between the leading and lagging strands at a replication fork. This mechanism is shown in Fig. 9B. It results in the same types of mutagenic events as illustrated above. (Adapted, with permission, from Sinden and Wells 1992.)

that was always longer than normal in individuals carrying the mutation (Oberle et al. 1991; Yu et al. 1991). A positive correlation was found between the link of the unstable sequence and the phenotype, since all affected individuals had one or more unstable restriction fragment bands with at least 600 bp more than the normal length. This instability was shown to be due to a variation in the copy number of a simple trinucleotide repeat $(CGG)_n$ (Fu et al. 1991; Kremer et al. 1991). Further analysis of the fragile X unstable element showed that normal individuals have approximately 5-54 copies of the triplet repeat, whereas carriers have 50-190 copies, and afflicted individuals have approximately 190-1000 copies (Fig. 6). The carrier copy number predisposes to further amplification in subsequent generations. This increase in copy number of the unstable sequence with subsequent generations accounts for a non-Mendelian genetic property, "anticipation," which is usually referred to in fragile X syndrome as the Sherman paradox (Verkerk et al. 1991; Sutherland et al. 1991; Caskey et al. 1992; Harley et al. 1992; Rennie 1992; Richards and Sutherland 1992). The penetrance of the disorder increases in successive generations of affected pedigrees, since there is a relationship between repeat copy number and phenotype (the copy number increases with successive generations).

The genetically unstable region contains multiple CGG repeats within the mRNA noncoding region of a fragile X associated gene (*FMR-1*). The CGG repeats are 5' to the translation initiation site of the *FMR-1* gene (Fu et al. 1991; Kremer et al. 1991; Pieretti et al. 1991; Yu et al. 1991). The CGG repeat is approximately 70 bp upstream of the AUG of the *FMR-1* gene, which is 592 amino acids in length, and a CG island is approximately 250 bp upstream of the triplet repeat; the function of the *FMR-1* gene is unknown.

Methylation appears to play an important role in the expression of this locus. In affected males with more than 200 copies of the repeat, there is no expression of the *FMR-1* gene, which may be the result of methylation of a CpG island several hundred base pairs 5' to the CGG repeats (Pieretti et al. 1991). Methylation of this CpG island associated with the *FMR-1* gene has been observed in fragile X patients (Bell et al. 1991).

Myotonic dystrophy

The myotonic muscular dystrophy gene shows molecular genetic similarities with the genetic arrangement described above. Myotonic dystrophy (DM) is an autosomal dominant disorder characterized primarily by myotonia and progressive muscle weakness, although central nervous system, cardiovascular, and ocular manifestations also are observed. The disease shows a marked variation in expression, ranging from a severe congenital form that is frequently fatal just after birth to

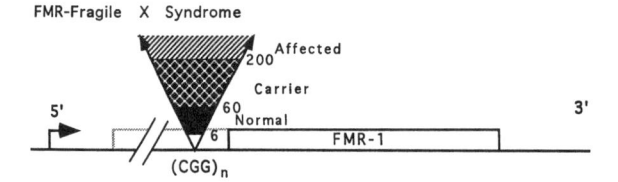

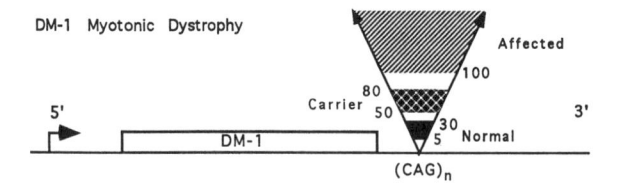

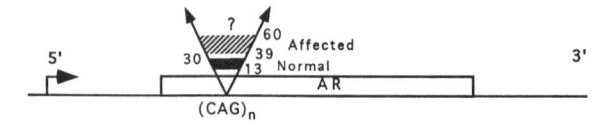

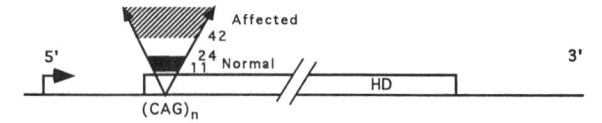

Figure 6 Molecular etiology of human genetic diseases involving expansion of a triplet repeat. In fragile X syndrome-associated gene *FMR-1*, there are ~29 copies of a CCG triplet repeat, 5′ to the translation start site within the noncoding region of the mRNA. In normal individuals there are ~5–54 copies of the triplet repeat. This expands to ~50–190 copies in carrier individuals and to ~190–1000 copies in affected individuals. In a case of muscular dystrophy, there is a CAG triplet repeat in a 3′ untranslated region of the *DM-1* gene. Normal individuals have 5–30 copies of the sequence, carriers 50–80 copies, and affected individuals 100–5000 copies. In Kennedy's disease, there is a CAG repeat within a gene encoding the androgen receptor. This triplet repeat encodes a run of glutamines. Normal individuals have 10–30 triplet repeats; affected individuals have 40–60 repeats. In Huntington's disease, a CAG triplet repeat is found within a gene *IT15*. It too is believed to encode a run of glutamines. In Huntington's disease, fewer than 42 repeats are found in normal individuals, whereas affected individuals have more than 42 triplet repeats. (Adapted, with permission, from Richards and Sutherland 1992, copyright by Cell Press.)

the asymptomatic form associated with normal longevity. DM in affected families may exhibit genetic anticipation (described above) (Brook et al. 1992; Caskey et al. 1992; Fu et al. 1992; Harley et al. 1992; Mahadevan et al. 1992; Tinsley et al. 1992). The DM mutation involves an unstable region (Brook et al. 1992; Fu et al. 1992; Harley et al. 1992; Mahadevan et al. 1992) and expansion of a $(CTG)_n$ repeat located in the 3'-untranslated region of the *DM-1* gene (methionine protein kinase) (Brook et al. 1992; Fu et al. 1992; Tinsley et al. 1992). Normal individuals have 5–30 copies of this sequence, whereas carriers have 50–80 copies. Affected individuals seem to have between 100 and 5000 copies of the triplet repeat (Fig. 6). As with the fragile X syndrome, the severity of the disease increases with longer repeat lengths. Genetic instability in somatic tissue is observed also in patients containing long CTG repeats (Richards and Sutherland 1992). DM is one of the most variable inherited disorders known in clinical medicine and is the most common muscular dystrophy of adults.

Kennedy's disease

Kennedy's disease (spino-bulbar muscular atrophy) is an X-linked recessive genetic disorder characterized by progressive muscular weakness of upper and lower extremities that starts in adults and is secondary to neural degeneration (Harding et al. 1982). This rare muscle disorder also involves a CTG repeat (LaSpada et al. 1991). The repeat is located within the gene for the androgen receptor (in exon 1), and the triplet CTG sequence encodes a tract of glutamines. Presumably, the enlarged glutamine tract interferes with the normal functioning of the androgen receptor. The deleterious consequences of including potentially unstable trinucleotide repeats within a gene make it logical to conclude that their presence in genes must have functional significance. Interestingly, polyglutamine tracts are frequently found in DNA-binding transcription factors such as SP1 and TFIID (Mitchell and Tjian 1989). Normal individuals have between 10 and 30 triplet repeats, whereas afflicted individuals have between 40 and 60 repeat units. DNA instability has not been observed, possibly due to the shorter sequences involved, which apparently do not exhibit large dynamic expansion mutations.

Huntington's disease

A very recent paper describes a novel gene containing a trinucleotide repeat (CTG) that is expanded and unstable on Huntington's disease (HD) chromosomes (MacDonald et al. 1993). The new gene, *IT15*, isolated using cloned, trapped exons from the target area, contains a polymorphic trinucleotide repeat that is expanded and unstable on HD chromosomes. The normal range of the triplet repeat length is less than

24 repeats. CTG tracts longer than 48 repeats were observed on HD chromosomes for all 75 diseased families examined, comprising a variety of ethnic backgrounds and haplotypes (Fig. 6). The CTG repeat appears to encode glutamine within the coding sequence of a predicted 348-kD protein that is widely expressed but unrelated to any known gene. Thus, the HD mutation involves an unstable DNA segment, similar to those described above, acting in the context of a novel gene to produce a dominant phenotype.

Anticipation

Several interesting review articles have appeared in the recent past concerning the impact of these "mutable mutations" on molecular biology and human genetics in clinical practice (Caskey et al. 1992; Davies 1992; Harper et al. 1992; Rennie 1992; Sutherland and Richards 1992; Yu et al. 1992). The discovery of a new mechanism of genome organization and function is very rare. The realization of a possible molecular basis for anticipation was dramatically revealed in a short time period rather than as the product of long and systematic investigations. Anticipation was legitimized by the discovery of expanded triplet repeat sequences as correlated with disease syndromes. Previously, anticipation was not accepted by geneticists, since the concept was strongly antieugenic, and it was believed that poor clinical diagnosis served as the explanation. Furthermore, the genome was considered to be stable, and no mechanism was known by which the virtually Lamarckian entity could operate.

MUTAGENESIS OF TRIPLET REPEATS: POSSIBLE MECHANISMS

There may be two different mutagenic events occurring at the DNA sequence level that give rise to the expansion of triplet repeats. The first type of mutagenic event is "small" expansion to a genetically unstable length. In the case of Kennedy's and Huntington's diseases, the expansion need only occur by a factor of 2–4 to cause the disease state. This is because the $(CAG)_n$ triplet repeat in both these diseases is believed to encode a polyglutamine tract in the protein, and expansion of this tract destroys the normal protein function. However, in Kennedy's and Huntington's diseases, if one considers the shortest normal allele and the longest allele from an affected individual, expansion may occur by a factor of 10. In the case of fragile X, the expansion occurs from the normal <60 repeats to a carrier range of 60–200 repeats. Following this 2- to 4-fold expansion, affected individuals frequently contain very long repeats that have expanded by more than a factor of 10.

The basic mutagenic mechanisms leading to the 2- to 4-fold and >10-fold expansion may be different. Therefore, for fragile X and DM, there may be a "carrier-type" mutation mechanism and an "affected-individual-type" mechanism. As discussed below, there are several mechanisms that could lead to carrier-type expansion by a factor of 2-3. How, in a single affected-individual-type mutational event, presumably during meiosis in the female chromosome, does the length of the triplet repeat expand by more than a factor of 10? There is no mutagenic mechanism known that can give rise to linear expansion of a region of the genome by a factor of 10. However, there are several possible mechanisms for massive triplet expansion, based on our understanding of the known potential mechanisms for mutagenesis and the potential for the formation of alternate DNA structures by triplet repeats. The following discussion attempts to summarize some of the possible models and ideas that might be tested experimentally.

Genetic recombination-based mutation

Experimental evidence from extrachromosomal studies in human cells has correlated the presence of simple sequence repeats with an elevated recombination frequency in extrachromosomal recombination (Wahls et al. 1990). Moreover, hot spots of meiotic recombination colocalize with simple sequence repeats in the H2 complex (Steinmetz et al. 1986). In mammalian cells, the minimal efficient processing segment (MEPS), below which recombination is very inefficient, is about 150-200 base pairs (Rubnitz and Subramani 1984; Liskay et al. 1987). This size corresponds with the length of triplet repeats when instability is first observed (the carrier stage). However, genetic recombination events leading to the new linkage of flanking markers have not been detected in affected individuals.

One mechanism that could account for both somatic and germ-line instability in repeat length invokes a specific cleavage of the chromosome and repair by a double strand break-recombination model (Fig. 7) (Szostak et al. 1983; Hastings 1992; Kobayashi 1992). Several steps are involved in expansion by a double strand break (or genetic recombination model). First, triplet repeats would be the site of frequent chromosome breaks, providing an initiation event in recombination (or repair). Exonucleolytic digestion would expose a single strand region. Strand synapsis during recombinational repair would occur over short regions of triplet repeats between two homologous chromosomes. DNA polymerase and ligase activity would synthesize an intact single strand on the chromosome containing the original break. The two chromosomes would separate (with assistance of topoisomerase activity), and repair synthesis would complete expansion of the chromosome containing the original break. Note that simple recombination alone cannot account for expansion by more than a factor of 2 or possibly 3 in a single recombination event. This type of event may lead to somatic variation or to expansion from a normal to a carrier stage. Multiple breaks and

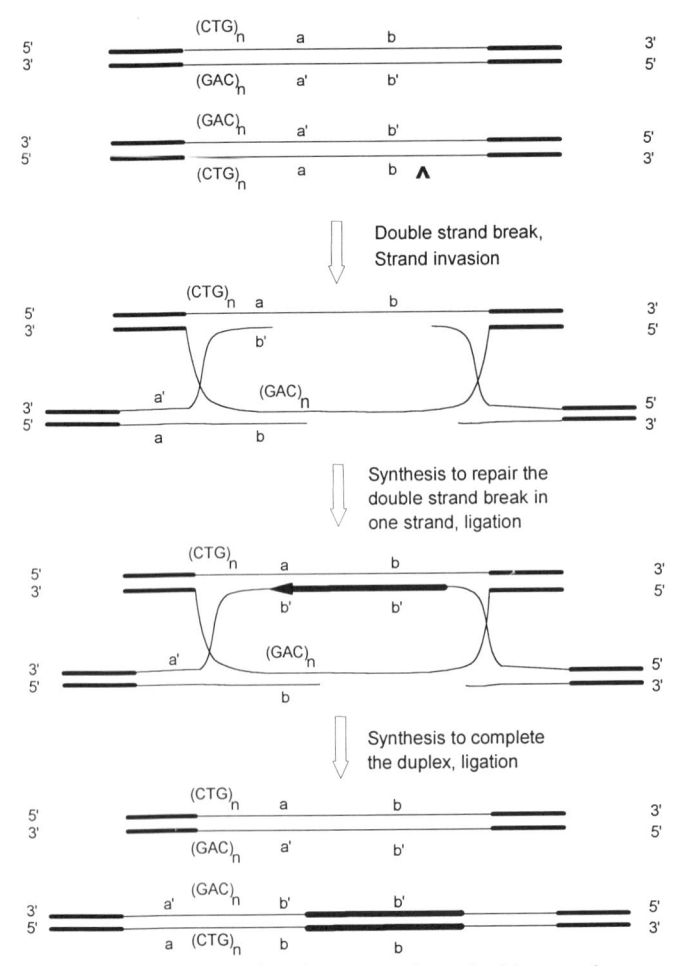

Figure 7 Expansion of triplet repeats by a double strand break mechanism. Following the alignment of two homologous chromosomes (*top*), if a double strand break occurs within a triplet repeat region (at the caret), exonucleolytic digestion can produce single strand ends on the duplex molecule that can invade and pair with the triplet region of the homologous chromosome. In the example shown, the double strand break is 3' to position b near the right end of the triplet repeat (shown by the thin lines). The b' region 5' end pairs with position a in the homologous chromosome, whereas the 3' OH end of the triplet repeat pairs to the region to the right of b (second structure). With this pairing of the breakpoints near the termini of the triplet repeat region, continued synthesis and ligation will result in expansion of the triplet repeat region (bold arrow, third structure). Following synthesis of an intact strand, the chromosomes can separate and DNA repair synthesis can fill in the resulting gap to complete a duplex chromosome (bottom structure).

recombinational events would be required for expansion by a factor of 10 or more, from a carrier to the affected individual stage. DNA amplification by an onion skin replication mechanism (Stark and Wahl 1984) followed by multiple genetic recombination events might also lead to massive expansion of triplet repeats.

DNA replication-based mutations

The triplet expansion may occur as a result of replication-based errors. Slipped mispairing (Streissinger et al. 1966) can account for duplications and deletions of single bases, units of repeating bases, or DNA between direct repeats. The lengths of di-, tri-, and tetranucleotide repeats in the human genome are known to vary (Weber 1990; Morral et al. 1991; Zielenski et al. 1991). Slippage during DNA replication can explain heterogeneity of repeats in the human population or within somatic cells of a single individual. Moreover, slippage could eventually give rise to the genetic change from a normal repeat length to a carrier size. A single slippage could not account for the massive amplification observed from a carrier length to a severely affected individual. To explain expansion by more than a factor of 10, repeated or multiple slippage events would have to occur (Fig. 8). DNA polymerase in a reiterative synthesis mode has been known to generate long simple repeated sequences thousands of base pairs in length from short primers (Kornberg et al. 1964; Wells et al. 1965, 1967). Slipped misalignment during DNA replication of many repeats can lead to the production of long polymers (Schlötterer and Tautz 1992). The rate of slippage is dependent on repeat length, complexity, and stability, but independent of the length of the progeny strand.

What could promote multiple slippage events? A strong block to replication might provide the opportunity for multiple slippage events by an "idling polymerase" (Fig. 8). A strong block might be provided by an alternate DNA secondary structure that is stabilized either by DNA supercoiling, by proteins, or by both. In studies involving the CGG repeats of fragile X and the CAG repeats of DM, the thermophilic DNA polymerases used for PCR amplification cannot get through long regions of the repeats (Kremer et al. 1991; Fu et al. 1991, 1992; Brook et al. 1992; Mahadevan et al. 1992). However, this may not involve slippage but the formation of complex secondary structures upon denaturation and annealing. It is known, for example, that a potential triplex region provides a block to replication in vitro and in vivo (Baran et al. 1991; Brinton et al. 1991). DNA polymerase may dissociate from the 3′ end, allowing breathing and potential slipped misalignment stabilized by base pairs in a hairpin-like stem. Both GGC and CTG triplet repeats could potentially form duplex DNA structures in which 2 out of 3 bases pair in a Watson-Crick form. Alternatively, the GGC repeat could form a parallel duplex

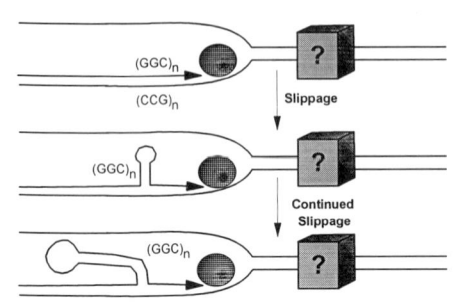

Figure 8 Reiterative synthesis resulting in the expansion of triplet repeats. DNA polymerase may stall or enter a reiterative synthesis mode resulting in the expansion of triplet repeats. The polymerase may become stalled due to a block in polymerase movement (*shaded box*). The block may be caused by the formation of an alternate secondary structure involving triplet repeats, a tightly bound DNA-binding protein, or the interaction of DNA with a macromolecular structure in the nucleus. Reiterative synthesis of DNA polymerase I on small templates has been shown to produce DNA polymers, thousands of base pairs in length. A displaced strand resulting from reiterative synthesis could form DNA secondary structures in the case of GGC triplet repeats and CTG triplet repeats in which every two out of three bases could form Watson-Crick base pairs. (Adapted, with permission, from Sinden and Wells 1992.)

with G:G base pairs similar to that found in telomere DNA (Sundquist and Klug 1989; Sen and Gilbert 1990).

Another replication-based model is one in which strand-displacement synthesis repeatedly occurs. Strand displacement and strand switch are known to occur especially during replication from a nick (Masamune and Richardson 1981; Lechner et al. 1983). Figure 9A shows several models for triplet expansion involving strand displacement. Replication from a nick can lead to polymerization and displacement of the strand ahead of the DNA polymerase. If an intermolecular strand switch occurred, continued replication would lead to a "panhandle" structure (Fig. 9, pathway a). Another possibility involves rehybridization of the displaced strand with concomitant displacement of the newly replicated strand. The formation of a hairpin at the end of the newly synthesized strand could lead to synthesis of a hairpin stem (an intramolecular strand switch) as shown in Figure 9, pathway b. Both of these strand-switch mechanisms following strand-displacement synthesis would lead to inversions of the triplet repeat sequence. A strand switch between the leading and lagging strand (Fig. 9B) could also lead to expansion and inversion of a region of the triplet repeats. DNA se-

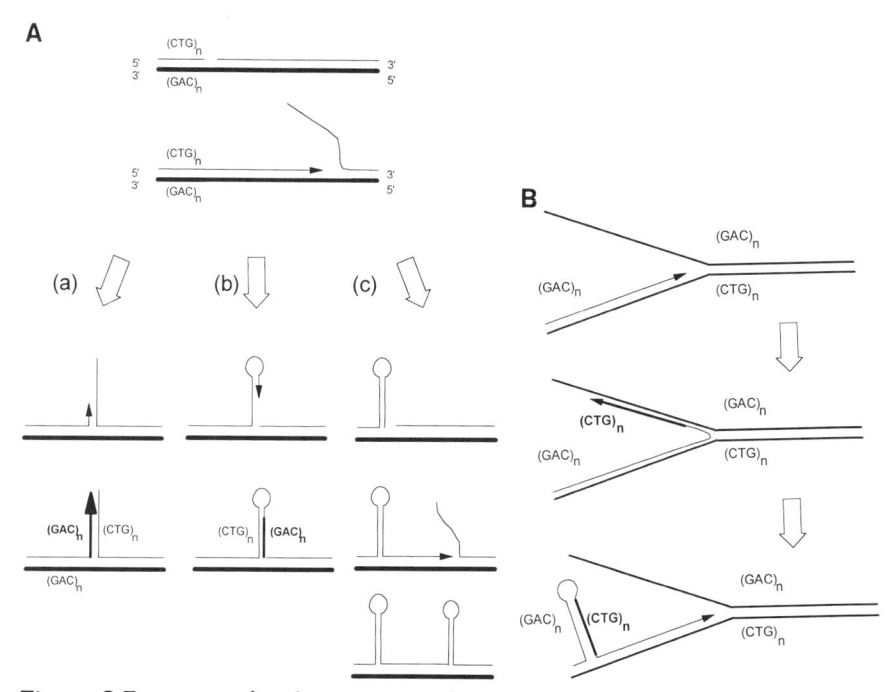

Figure 9 Expansion of triplet repeats involving strand switch mechanisms. (A) Expansion of triplet repeats involving strand-displacement synthesis is shown. Following the introduction of a nick in one strand of a triplet repeat (*top*), DNA synthesis without 5′ to 3′ exonuclease digestion results in displacement of the strand ahead of the DNA polymerase (second structure). The $(CTG)_n$ strand is represented by the thin line and the $(GAC)_n$ strand by the thick line. In pathway a, an intermolecular strand switch occurs in which polymerase copies up the displaced strand, resulting in a panhandle structure. The thick line in the top strand represents a $(GAC)_n$ region templated off the $(CTG)_n$ displaced strand. In pathway b, an intramolecular strand switch occurs following dissociation of the newly synthesized strand (by branch migration of the displaced strand) and the formation of a hairpin loop that primes synthesis down the stem. In pathway c, following strand-displacement synthesis, the displaced strand branch migrates back into a duplex, pairing with the template strand. The newly synthesized strand forms a DNA secondary structure positioning the 3′ end on the template, allowing the triplet repeat region to be recopied. Eventual ligation results in covalent closure of the newly synthesized strand. In pathways a and b, an inversion of a region of the triplet repeat would occur. In pathway c, no inversion of the triplet repeat sequence would occur. (B) Intermolecular strand switch occurs at a replication fork. Following the strand switch, synthesis occurs on the opposite strand, resulting in an inversion of a region of the triplet repeat. This produces an inverted repeat. The newly synthesized strand can become dissociated, form a hairpin structure, and repair with its original template strand. Continued synthesis from this hairpin structure would result in the expansion (and inversion) of a triplet repeat.

quence analysis (as far as is possible) does not reveal inversion events (T.P. Yang, unpubl.). Moreover, these strand-displacement/strand switch models would only lead to expansion by a factor of 2.

The strand-displacement/slippage model shown in Figure 9, pathway c, could lead to expansion by unit lengths *without* inversion. After strand-displacement synthesis occurs, the displaced strand folds into a hairpin arm in which two of every three bases in a GGC or CTG triplet repeat could base-pair in a Watson-Crick manner. The third base of the triplet would form GG, CC, AA, or TT mismatches, which in some cases have been shown to occur (Arnold et al. 1987; Skelly et al. 1993). The newly synthesized strand would then ligate to the 5'end of the displaced quasi-hairpin stabilized strand. A second nick could be introduced, and a second round of strand displacement could occur. Repeated rounds of strand displacement and ligation could lead to expansion.

Acknowledgments

Work from the authors' laboratories is supported by grants from the National Institutes of Health (GM-30822), from the National Science Foundation (86-07785), and from the Robert A. Welch Foundation to R.D.W. and by grants from National Institutes of Health (GM-37677) and the National Institute of Environmental Health and Safety (ES-05508 and P-01ES-05652) to R.R.S. We thank W.A. Rosché and S. Kang for assistance in preparation of the figures.

References

Arnold, F.H., S. Wolk, P. Cruz, and I. Tinoco, Jr. 1987. Structure, dynamics, and thermodynamics of mismatched DNA oligonucleotide duplexes d(CCCAGGG) 2 and d(CCCTGGG) 2. *Biochemistry* 26: 4068.

Baran, N., A. Lapidot, and H. Manor. 1991. Formation of DNA triplexes accounts for arrests of DNA synthesis at d(TC)$_n$ and d(GA)$_n$ tracts. *Proc. Natl. Acad. Sci.* 88: 507.

Bebenek, K., J. Abbotts, J.D. Roberts, S.H. Wilson, and T.A. Kunkel. 1989. Specificity and mechanism of error-prone replication by human immunodeficiency virus-1 reverse transcriptase. *J. Biol. Chem.* 264: 16948.

Bebenek, K., J. Abbotts, S.H. Wilson, and T.A. Kunkel. 1993. Error-prone polymerization by HIV-1 reverse transcriptase: Contribution of template-primer misalignment, miscoding, and termination probability to mutational hotspots. *J. Biol. Chem.* 268: 10324.

Bell, M.V., M.C. Hirst, Y. Nakahori, R.N. MacKinnon, A. Roche, T.J. Flint, and P.A. Jacobs. 1991. Physical mapping across the fragile X: Hypermutation and clinical expression of the fragile X syndrome. *Cell* 64: 861.

Brinton, B.T., M.S. Caddle, and N.H. Heintz. 1991. Position and orientation-dependent effects of a eukaryotic Z-triplex DNA motif on episomal DNA replication in COS-7 cells. *J. Biol. Chem.* **266**: 5153.

Brook, J.D., M.E. McCurrach, H.G. Harley, A.J. Buckler, D. Church, H. Aburatani, K. Hunter, V.P. Stanton, J.P., Thirion, T. Hudson, R. Sohn, B. Zemelman, R.G. Snell, S.A. Rundle, S. Crow, J. Davies, P. Shelbourne, J. Buxton, C. Jones, V. Juvonen, K. Johnson, P.S. Harper, D.J. Shaw, and D.E. Housman. 1992. Molecular basis of myotonic dystrophy: Expansion of a trinucleotide (CTG) repeat at the 3' end of a transcript encoding a protein kinase family member. *Cell* **68**: 799.

Bruckner, R.C., R.E. Dutch, B. Zemelman, E.S. Mocarski, and I.R. Lehman. 1992. Recombination *in vitro* between herpes simplex virus type 1a sequences. *Proc. Natl. Acad. Sci.* **89**: 10950.

Caskey, C.T., A. Pizzuti, Y.-H. Fu, R.G. Fenwick, Jr., and D. Nelson. 1992. Triplet repeat mutations in human disease. *Science* **256**: 784.

Collins, J. 1980. Instability of palindromic DNA in *Escherichia coli*. *Cold Spring Harbor Symp. Quant. Biol.* **45**: 409.

Creighton, S., M. Huang, H. Cai, N. Arnheim, and M.F. Goodman. 1992. Base mispair extension kinetics. *J. Biol. Chem.* **267**: 2633.

Dasgupta, U., K. Weston-Hafer, and D.E. Berg. 1987. Local DNA sequence control of deletion formation in *Escherichia coli* plasmid pBR322. *Genetics* **115**: 41.

Davies, K. 1992. Too many trinucleotide repeats. *Nature* **358**: 90.

Drake, J.W., B.W. Glickman, and L.S. Ripley. 1983. Updating the theory of mutation. *Am. Sci.* **71**: 621.

Edwards, A., H.A. Hammond, L. Jin, C.T. Caskey, and R. Chakraborty. 1992. Genetic variation at five trimeric and tetrameric tandem repeat loci in four human population groups. *Genomics* **12**: 241.

Farabaugh, P.J., U. Schmeissner, M. Hofer, and J.H. Miller. 1978. Genetic studies of the lac repressor on the molecular nature of spontaneous hotspots in the LacI gene of *Escherichia coli*. *J. Mol. Biol.* **126**: 847.

Freund, A.M., M. Bichara, and R.P. Fuchs. 1989. Z-DNA-forming sequences are spontaneous deletion hot spots. *Proc. Natl. Acad. Sci.* **86**: 7465.

Fu, Y.-H., D.P.A. Kuhl, A. Pizzuti, M. Pieretti, J.S. Sutcliffe, S. Richards, A.J.M.H. Verkerk, J.J.A. Holden, R.G. Fenwick, Jr., S.T. Warren, B.A. Oostra, D.L. Nelson, and C.T. Caskey. 1991. Variation of the CGG repeat at the fragile X site results in genetic instability: Resolution of the sherman paradox. *Cell* **67**: 1047.

Fu, Y.-H., A. Pizzuti, R.G. Fenwick, Jr., J. King, S. Rajnarayan, P.W. Dunne, J. Dubel, G.A. Nasser, T. Ashizawa, P. DeJong, B. Wieringa, R. Korneluk, M.B. Perryman, H.F. Epstein, and C.T. Caskey. 1992. An unstable triplet repeat in a gene related to myotonic muscular dystrophy. *Science* **255**: 1256.

Glickman, B.W. and L.S. Ripley. 1984. Structural intermediates of deletion mutagenesis: A role for palindromic DNA. *Proc. Natl. Acad. Sci.* **81**: 512.

Goodman, S.D., S.C. Nicholson, H.A. Nash. 1992. Deformation of DNA during site-specific recombination of bacteriophage λ: Replacement of IHF protein by HU protein or sequence-directed bends. *Proc. Natl. Acad. Sci.* **89**: 11910.

Hagerman, P.J. 1990. Sequence-directed curvature of DNA. *Annu. Rev. Biochem.* **59**: 755.

Harding, A.E., P.K. Thomas, M. Baraitser, P.G. Bradbury, J.A. Morgan-Hughes, and J.R. Ponsford. 1982. X-linked recessive bulbospinal neuronopathy: A report of ten cases. *J. Neurol. Neurosurg. Psychiatry* **45**: 1012.

Harley, H.G., S.A. Rundle, W. Reardon, J. Myring, S. Crow, J.D. Brook, P.S. Harper, and D.J. Shaw. 1992. Unstable DNA sequence in myotonic dystrophy. *Lancet* **339**: 1125.

Harper, P.S., H.G. Harley, W. Reardon, and D. Shaw. 1992. Anticipation in myotonic dystrophy: New light on an old problem. *Am. J. Hum. Genet.* **51**: 10.

Hartman, D., K. Shu-Ru, T. Broker, L.T. Chow, and R.D. Wells. 1992. Intermolecular triplex formation distorts the DNA duplex in the regulatory region of human papillomavirus type-11. *J. Biol. Chem.* **267**: 5488.

Hastings, P.J. 1992. Mechanism and control of recombination in fungi. *Mutat. Res.* **284**: 97.

Jaworski, A., J.A. Blaho, J.E. Larson, M. Shimizu, and R.B. Wells. 1989. Tetracycline promoter mutations decrease non-B DNA structural transitions, negative linking differences and deletions in recombinant plasmids in *Escherichia coli*. *J. Mol. Biol.* **207**: 513.

Jaworski, A., W.T. Hsieh, J.A. Blaho, J.E. Larson, and R.D. Wells. 1987. Left-handed DNA *in vivo*. *Science* **238**: 773.

Joyce, C.M., X.C. Sun, and N.D.F. Grindley. 1992. Reactions at the polymerase active site that contribute to the fidelity of *Escherichia coli* DNA polymerase I (Klenow Fragment). *J. Biol. Chem.* **267**: 24485.

Kang, S. and R.D. Wells. 1992. Central non-pur pyr sequences in oligo dG dC tracts and metal ions influence the formation of intramolecular DNA triplex isomers. *J. Biol. Chem.* **26**: 20887.

Kang, C., X. Zhang, R. Ratliff, R. Moyzis, and A. Rich. 1992. Crystal sructure of four-stranded Oxytricha telomeric DNA. *Nature* **356**: 126.

Klysik, J. 1992. Cruciform extrusion facilitates intramolecular triplex formation between distal oligopurine·oligopyrimidine tracts: Long range effects. *J. Biol. Chem.* **267**: 17430.

Kobayashi, I. 1992. Mechanisms for gene conversion and homologous recombination: The double-strand break repair model and the successive half crossing-over model. *Adv. Biophys.* **28**: 81.

Kohwi-Shigematsu, T. and Y. Kohwi. 1991. Detection of triple-helix related structures adopted by poly(dG)-poly(dC) sequences in supercoiled plasmid DNA. *Nucleic Acids Res.* **19**: 4267.

Kornberg, A., L.L. Bertsch, J.F. Jackson, and H.G. Khorana. 1964. Enzymatic synthesis of deoxyribonucleic acid, XVI. Oligonucleotides as templates and the mechanism of their replication. *Proc. Natl. Acad. Sci.* **51**: 315.

Krawczak, M. and D.N. Cooper. 1991. Gene deletions causing human genetic disease: Mechanisms of mutagenesis and the role of the local DNA sequence environment. *Hum. Genet.* **86**: 425.

Kremer, E.J., M. Pritchard, M. Lynch, S. Yu, K. Holman, E. Baker, S.T. Warren, D. Schlessinger, G.R. Sutherland, and R.I. Richards. 1991. Mapping of DNA instability at the fragile X to a trinucleotide repeat sequence p(CCG)$_n$. *Science* **252**: 1711.

Kunkel, T.A. 1990. Misalignment-mediated DNA synthesis errors. *Biochemistry* **29**: 8003.

———. 1992. DNA replication fidelity. *J. Biol. Chem.* **267**: 18251.

La Spada, F.M., E.M. Wilson, D.B. Lubahn, A.E. Harding, K.H. Fischbeck. 1991. Androgen receptor gene mutations in X-linked spinal and bulbar muscular atrophy. *Nature* **352**: 77.

Lechner, R.L., M.J. Engler, and C.C. Richardson. 1983. Characterization of strand displacement synthesis catalyzed by bacteriophage T7 DNA polymerase. *J. Biol. Chem.* **258**: 11174.

Levinson, G. and G.A. Gutman. 1987. Slipped-strand mispairing: A major mechanism for DNA sequence evolution. *Mol. Biol. Evol.* **4**: 203.

Liskay, R.M., A. Letsou, and J.L. Stachelek. 1987. Homology requirement for efficient gene conversation between duplicated chromosomal sequences in mammalian cells. *Genetics* **115**: 161.

MacDonald, M.E. And 57 other authors. 1993. A novel gene containing a trinucleotide repeat that is expanded and unstable on Huntington's Disease chromosomes. *Cell* **72**: 971.

Mahadevan, M., C. Tsilfidis, L. Sabourin, G. Shutler, C. Amemiya, G. Jansen, C. Neville, M. Narang, J. Barcelo, K. O'Hoy, S. Leblond, J.E. McDonald, P.J. de Jons, B. Wieringa, and R.G. Korneluk. 1992. Myotonic dystrophy mutation: An unstable CTG repeat in the 3' untranslated region of the gene. *Science* **255**: 1253.

Masamune, Y. and C.C. Richardson. 1981. Strand displacement during deoxyribonucleic acid synthesis at single strand breaks. *J. Biol. Chem.* **246**: 2692.

McClellan, J.A., P. Boublikova, E. Palecek, and D.M. Lilley. 1990. Superhelical torsion in cellular DNA responds directly to environmental and genetic factors. *Proc. Natl. Acad. Sci.* **87**: 8373.

McKeon, C., A. Schmidt, and B. deCrombrugghe. 1984. A sequence conserved in both the chicken and mouse a2(I) collagen promoter contains sites sensitive to S1 nuclease. *J. Biol. Chem.* **259**: 6636.

Mitchell, P.J. and R. Tjian. 1989. Transcription regulation in mammalian cells by sequence-specific DNA binding proteins. *Science* **245**: 371.

Morral, N., V. Nunes, T. Casals, and X. Estivill. 1991. CA/GT microsatellite alleles within the cystic fibrosis transmembrane conductance regulator (CFTR) gene are not generated by unequal crossingover. *Genomics* **10**: 692.

Murchie, A.I.H., R. Bowater, F. Aboul-Ela, and D.M.J. Lilley. 1992. Helix opening transitions in supercoiled DNA. *Biochem. Biophys. Acta* **1131**: 1.

Noirot, P., J. Bargonetti, and R.P. Novick. 1990. Initiation of rolling-circle replication in pT181 plasmid: Initiator protein enhances cruciform extrusion at the origin. *Proc. Natl. Acad. Sci.* **87**: 8560.

Oberle, I., F. Rousseau, D. Heitz, C. Kretz, D. Devys, A. Hanauer, J. Boue, M.F. Bertheas, and J.L. Mandel. 1991. Instability of a 550-base pair DNA segment and abnormal methylation in fragile X syndrome. *Science* **252**: 1097.

Ohshima, A., S. Inouye, and M. Inouye. 1992. *In vivo* duplication of genetic elements by the formation of stem-loop DNA without an RNA intermediate. *Proc. Natl. Acad. Sci.* **89**: 1016.

Panyutin, I.G., O.I. Kovalsky, E.I. Budowsky, R.E. Dickerson, M.E. Rikhirev, and A.A. Lipanov. 1990. G-DNA: A twice-folded DNA structure adopted by single-stranded oligo(dG) and its implications for telomeres. *Proc. Natl. Acad. Sci.* **87**: 867.

Panyutin, I. and R.D. Wells. 1992. Nodule DNA in the $(GA)_{37} \cdot (CT)_{37}$ insert in su-

perhelical plasmids. *J. Biol. Chem.* **267:** 5495.
Papanicolaou, C. and L.S. Ripley. 1991. An in vitro approach to identifying specificity determinants of mutagenesis mediated by DNA misalignments. *J. Mol. Biol.* **221:** 805.
Pierce, J.C., D. Kong, and W. Masker. 1991. The effect of the length of direct repeats and the presence of palindromes on deletion between directly repeated DNA sequences in bacteriophage T7. *Nucleic Acids Res.* **14:** 3901.
Pieretti, M., F. Zhang, Y.-H. Fu, S.T. Warren, B.A. Oostra, C.T. Caskey, and D.L. Nelson. 1991. Absence of expression of the FMR-1 gene in fragile X syndrome. *Cell* **66:** 817.
Raghuraman, M.K. and T.R. Cech. 1990. Effect of monovalent cation-induced telomeric DNA structure on the binding of oxytricha telomeric protein. *Nucleic Acids Res.* **18:** 4543.
Rahmouni, A.R. and R.D. Wells. 1992. Direct evidence for the effect of transcription on local DNA supercoiling *in vivo*. *J. Mol. Biol.* **223:** 131.
Rennie, J. 1992. Mutable mutation. *Sci. Am.* **266:** 34.
Rich, A., A. Nordheim, and A.H.-J. Wang. 1984. The chemistry and biology of left-handed-DNA. *Annu. Rev. Biochem.* **53:** 791.
Richards, R.I. and G.R. Sutherland. 1992. Heritable unstable DNA sequences and human genetic disease. *Cell* **70:** 709.
Ripley, L.S. 1982. Model for the participation of quasi-palindromic DNA sequences in frameshift mutation. *Proc. Natl. Acad. Sci.* **79:** 4128.
⸺. 1991. Frameshift mutation: Determinants of specificity. *Annu. Rev. Genet.* **24:** 189.
Robertson, C.A. and H.A. Nash. 1988. Bending of the bacteriophage λ attachment site by *Escherichia coli* integration host factor. *J. Biol. Chem.* **263:** 3554.
Rubnitz, J. and S. Subramani. 1984. The minimum amount of homology required for homologous recombination in mammalian cells. *Mol. Cell Biol.* **5:** 529.
Schlötterer, C. and D. Tautz. 1992. Slippage synthesis of simple sequence DNA. *Nucleic Acid Res.* **20:** 211.
Sen, D. and W. Gilbert. 1990. A sodium-potassium switch in the formation of four-stranded G4-DNA. *Nature* **344:** 410.
Shimizu, M., J. Hanvey, and R.D. Wells. 1990. Multiple non-B-DNA conformations of polypurine polypyrimidine sequences in plasmids. *Biochemistry* **29:** 4704.
Sinden, R.R. and R.D. Wells. 1992. DNA structure, mutations, and human genetic disease. *Curr. Opin. Biotechnol.* **3:** 612.
Sinden, R.R., G. Zheng, R.G. Brankamp, and K.N. Allen. 1991. On the deletion of inverted repeated DNA in *Escherichia coli*: Effects of length, thermal stability, and cruciform formation *in vivo*. *Genetics* **129:** 991.
Skelly, J.V., K.J. Edwards, T.C. Jenkins, and S. Neidle. 1993. Crystal structure of an oligonucleotide duplex containing G·G base pairs: Influence of mispairing on DNA backbone conformation. *Proc. Natl. Acad. Sci.* **90:** 804.
Sowers, L.C., G.V. Fazakerley, R. Eritja, B.E. Kaplan, and M.F. Goodman. 1986. Base pairing and mutagenesis: Observations of a protonated base pair between 2-aminopurine and cytosine in an oligonucleotide by proton NMR. *Proc. Natl. Acad. Sci.* **83:** 5434.
Sowers, L.C., M.F. Goodman, R. Eritja, B.E. Kaplan, and G.V. Fazakerley. 1989.

Ionized and wobble base-pairing for bromouracil-guanine in equilibrium under physiological conditions: A nuclear magnetic resonance study on an oligonucleotide containing a bromouracil-guanine base-pair as a function of pH. *J. Mol. Biol.* **205:** 437.

Stark, G.R. and G.M. Wahl. 1984. Gene amplification. *Annu. Rev. Biochem.* **53:** 447.

Steinmetz, M., D. Stephan, and K.F. Lindahl. 1986. Gene organization and recombinational hotspots in the murine histocompatibility complex. *Cell* **44:** 895.

Stenzel, T.T., P. Patel, and D. Bastia. 1987. The integration host factor of *Escherichia coli* binds to bent DNA at the origin of replication of the plasmid pSC101. *Cell* **49:** 709.

Streisinger, G. and J. Owen. 1985. Mechanisms of spontaneous and induced frameshift mutation in bacteriophage T4. *Genetics* **109:** 633.

Streisinger, G., Y. Okada, J. Emrich, J. Newton, A. Tsugita, E. Terzaghi, and M. Inouye. 1966. Frameshift mutations and the genetic code. *Cold Spring Harbor Symp. Quant. Biol.* **31:** 77.

Sundquist, W.I. and A. Klug. 1989. Telomeric DNA dimerizes by formation of guanine tetrads between hairpin loops. *Nature* **342:** 825.

Sutherland, G.R. and R.I. Richards. 1992. Anticipation legitimized: Unstable DNA to the rescue. *Hum. Genet.* **50:** 238.

Sutherland, G.R., E.A. Haan, E. Kremer, M. Lynch, M. Pritchard, S. Yu, and R.I. Richards. 1991. Hereditary unstable DNA: A new explanation for some old genetic questions? *Lancet* **338:** 289.

Szostak, J.W., T.L. Orr-Weaver, R.J. Rothstein, and F.W. Stahl. 1983. The double-strand-break repair model of recombination. *Cell* **33:** 25.

Tinsley, J.M., D.J. Blake, A. Roche, U. Fairbrother, J. Riss, B.C. Byth, A.E. Knight, J. Kendrick-Jones, G.K. Suthers, D.R. Love, Y.H. Edwards, and K.E. Davies. 1992. Primary structure of dystrophin-related protein. *Nature* **360:** 591.

Trinh, T.Q. and R.R. Sinden. 1991. Preferential DNA secondary structure mutagenesis in the lagging strand of replication in *E. coli*. *Nature* **352:** 544.

———. 1993. The influence of primary and secondary DNA structure in deletion and duplication between direct repeats in *Escherichia coli*. *Genetics* **134:** 409.

Ulrich, M.J., W.J. Gray, and T.J. Ley. 1992. An intramolecular DNA triplex is disrupted by point mutations associated with hereditary persistence of fetal hemoglobin. *J. Biol. Chem.* **267:** 18649.

Ussery, D.W. and R.R. Sinden. 1993. Environmental influences on the *in vivo* level of intramolecular triplex DNA in *Escherichia coli*. *Biochemistry* **32:** 6206.

Verkerk, A.J.M.H., J. Pieretti, J.S. Sutcliffe, Y.-H. Fu, D.P.A. Kuhl, A. Pizzuti, O. Reiner, S. Richards, M.F. Victoria, F. Zhang, B.E. Eussen, G.-H.B. van Ommen, L.A.J. Blonden, G.J. Riggins, J.L. Chastain, C.B. Kunst, H. Galjaard, C.T. Caskey, D.L. Nelson, B.A. Oostra, and S.T. Warren. 1991. Identification of a gene (FMR-1) containing a CGG repeat coincident with a breakpoint cluster region exhibiting length variation in fragile X syndrome. *Cell* **65:** 905.

Wahls, W.P., L.J. Wallace, and P.D. Moore. 1990. The Z-DNA motif d(TG)30 promotes reception of information during gene conversion events while

stimulating homologous recombination in human cells in culture. *Mol. Cell Biol.* **10:** 785.
Ward, G.K., R. McKenzie, M. Zannis-Hadjopoulos, and G.B. Price. 1990. The dynamic distribution and quantification of DNA cruciforms in eukaryotic nuclei. *Exp. Cell Res.* **188:** 235.
Weber, J.L. 1990. Informativeness of human (dC-dA) n.(dG-dT) n polymorphisms. *Genomics* **7:** 524.
Wells, R.D. 1988. Unusual DNA structures. *J. Biol. Chem.* **263:** 1095.
Wells, R.D., E. Ohtsuka, and H.G. Khorana. 1965. Studies on polynucleotides L. Synthetic deoxyribopolynucleotides as templates for the DNA polymerase of *Escherichia coli*: A new double-stranded DNA-like polymer containing repeating dinucleotide sequences. *J. Mol. Biol.* **14:** 221.
Wells, R.D., T.M. Jacob, S.A. Narang, and H.G. Khorana. 1967. Studies on polynucleotides. LXIX. Synthetic deoxyribopolynucleotides as templates for the DNA polymerase of *Escherichia coli*: DNA-like polymers containing repeating trinucleotide sequences. *J. Mol. Biol.* **27:** 237.
Wells, R.D., D.A. Collier, J.C. Hanvey, M. Shimizu, and F. Wohlrab. 1988. The chemistry and biology of unusual DNA structures adopted by oligopurine·oligopyrimidine sequences. *FASEB J.* **2:** 2939.
Weston-Hafer, K. and D.E. Berg. 1991. Limits to the role of palindromy in deletion formation. *J. Bacteriol.* **173:** 315.
———. 1989. Palindromy and the location of deletion endpoints in *Escherichia coli*. *Genetics* **121:** 651.
Wohlrab, F. and R.D. Wells. 1989. Slight changes in conditions influence the family of non-B-DNA conformations of the Herpes Simplex Virus Type 1 DR2 repeats. *J. Biol. Chem.* **264:** 8207.
Wohlrab, F., S. Chatturjee, and R.D. Wells. 1991. The Herpes Simplex Virus I segment inversion site is specifically cleaved by a virus-induced nuclear endonuclease. *Proc. Natl. Acad. Sci.* **88:** 6432.
Wohlrab, F., M. McLean, and R.D. Wells. 1987. The segment inversion site of Herpes Simplex Virus Type 1 adopts a novel DNA structure. *J. Biol. Chem.* **262:** 6407.
Yu, S., M. Pritchard, E. Kremer, M. Lynch, J. Nancarrow, E. Baker, K. Homan, J.C. Mulley, S.T. Warren, D. Schlessinger, G.R. Sutherland, and R.I. Richards. 1991. Fragile X genotype characterized by an unstable region of DNA. *Science* **252:** 1179.
Yu, S., J. Mulley, D. Loesch, G. Turner, A. Donnelly, A. Gedeon, D. Hillen, E. Kremer, M. Lynch, M. Pritchard, G.R. Sutherland, and R.I. Richards. 1992. Fragile-X syndrome: Unique genetics of the heritable unstable element. *Hum. Genet.* **50:** 968.
Zheng, G., T. Kochel, R.W. Hoepfner, S.E. Timmons, and R.R. Sinden. 1991. Torsionally tuned cruciform and Z-DNA probes for measuring unrestrained supercoiling at specific sites in DNA of living cells. *J. Mol. Biol.* **221:** 107.
Zielenski, J., D. Markiewicz, F. Rininsland, J. Rommens, and L.-C. Tsui. 1991. A cluster of highly polymorphic dinucleotide repeats in intron 17b of the cystic fibrosis transmembrane conductance regulator (CFTR) gene. *Am. J. Hum. Genet.* **49:** 1256.

Homologous DNA Interactions in the Evolution of Gene and Chromosome Structure

Miroslav Radman,[1] Robert Wagner,[1] and Maja C. Kricker[2]

[1]Laboratoire de Mutagenese
Institute Jacques Monod
75251-Paris Cedex 05, France

[2]Computer Sciences Corporation
Durham, North Carolina 27713

The stability of mammalian genomes depends on the avoidance of recombination between repeated sequences such as transposable elements and pseudogenes. It appears that the structure of mammalian genes and chromosomes is a direct result of the need to avoid all homologous DNA interactions. Therefore, evolution seems to have strongly selected against extensive regions of perfect homology, both by reducing the size of homologous regions (by fragmenting coding regions into short exons) and by decreasing the degree of homology (by sequence divergence, especially in noncoding regions).

Points discussed include:

❏ utility and liability of recombination between homologous DNA sequences

❏ control of interrepeat recombination by the extent and degree of homology

❏ editing of homologous interactions in genetic recombination by the components of the mismatch repair system

❏ homologous DNA interactions other than recombination: RIP (repeat-induced point mutation) and MIP (methylation induced premeiotically)

❏ diversity of CpG dinucleotide frequencies in different genomic elements suggesting duplication-targeted CpG to TpG mutagenesis

❏ whether the fragmented status of mammalian genes into short exons protects coding sequences from deleterious homologous interactions with their processed pseudogenes

❏ the possible existence of a "hidden genetics" of interactions between some pseudogenes and their progenitor genes

❏ cases from human genetics

❏ sequence polymorphism as a key element of genomic stability and of neutral evolution: a role for "junk" DNA

INTRODUCTION

Homologous recombination is vitally important for at least two biological processes: (1) the repair of DNA damage, in particular double strand breaks, and (2) chromosomal disjunction in meiosis which requires pairing and crossover between homologs (for review, see Roeder 1990; Hawley and Arbel 1993; Radman and Wagner 1993). DNA repair can be accomplished by precise sister chromatid exchange, which has no deleterious genetic consequence. However, all other recombination between homologous DNA sequences, especially ectopic recombination between dispersed repeated DNA sequences resulting in chromosomal rearrangements, is potentially lethal. Thus, the higher the number of repeated sequences within a genome, the higher the risk of deleterious chromosomal rearrangements resulting from interrepeat recombination. In fact, it has been found that, in yeast meiosis, ectopic gene copies integrated in nonhomologous chromosomes recombine as efficiently as homologous genes in identical locations on homologous chromosomes (Haber et al. 1991).

Although the existence of numerous deletions resulting from recombination between abundant repetitive sequences such as mammalian short and long interspersed nuclear elements (SINEs and LINEs) has indicated the need to avoid recombination between such sequences (for review, see Hu and Worton 1992), little attention has been paid to the genetic risks of recombination between genes and pseudogenes (Kricker et al. 1992). However, such interactions can cause gene inactivation with and without the accompanying chromosomal rearrangements. There may be a "hidden genetics" of homologous interactions between genes and their pseudogenes, which may have a significant impact on human health, but which has not been actively studied because of the lack of interest in inactive pseudogenes, heretofore considered to be totally inert, "junk" DNA. One important reason for the complete sequencing of the human genome is to acquire knowledge of the number, structure, location, and sequence of pseudogenes and their relationship to human genetic diseases.

CONTROL OF INTERREPEAT RECOMBINATION BY LIMITING THE EXTENT AND DEGREE OF HOMOLOGY

The results of experiments with bacteria, yeast, *Drosophila*, plants, and mammalian cells indicate that there are common strategies, conserved throughout evolution, for the selective prevention of interrepeat recombination and mitotic recombination between homologs (which can lead to the expression of recessive mutations). The strategies have been to reduce both the extent and degree of homology between the potential partners in recombination (for review, see Radman 1988; Radman and Wagner 1993; Nassif and Engels 1993).

Recombination between cloned identical sequence repeats of variable length integrated ectopically into the same or different chromosomes (i.e., embedded in nonhomologous sequences) decreases roughly linearly with the decrease in their length down to a minimum size required for efficient recombination, below which recombination frequency decreases sharply. Extrapolation of the linear portion of the curve to zero recombination defines a minimum efficient processing segment (MEPS; Shen and Huang 1986). MEPSs have been found to be 25–30 bp long in *Escherichia coli* (Shen and Huang 1986), about 250 bp in yeast (Jinks-Robertson and Michelitch 1993), and 250–400 bp in plant cells (Puchta and Hohn 1991) and in cultured mammalian cells (for review, see Bollag et al. 1989). The fact that the length of MEPS is significantly less than the length of the heteroduplex joint region in homologous recombination (often several kilobases long) suggests that recombination enzymes are highly selective for sequence identity only at the initial stage of the strand exchange process. Once initiated, the strand exchange process can proceed even through large heterologies and form looped hetroduplex DNA structures (Lichten and Fox 1984). Therefore, under circumstances where there are extensive perfectly identical sequences, the interruption and fragmentation of large homologies by blocks of heterology is not an efficient means of recombination avoidance unless the fragmentation is into blocks less than MEPS length. A decrease in the degree, rather than the extent of DNA homology (e.g., by well-distributed point mutation sequence divergence) is more efficient in preventing homologous recombination, since sequence divergence of only a few percent decreases recombination by orders of magnitude (for review, see Radman 1988; Radman and Wagner 1993). For example, in yeast meiosis, efficient recombination occurs between blocks of perfect homology embedded in nonhomologous regions, whereas random divergence of as little as 10% prevents recombination between entire "homoeologous" chromosomes (Resnick et al. 1989; Haber et al. 1991).

The basic structure of mammalian genes and chromosomes suggests that both strategies for recombination avoidance are employed. In coding sequences, the MEPS-limitation strategy, i.e., fragmentation into shorter-than-MEPS exons, is employed, whereas in noncoding DNA sequences, the divergence/polymorphism strategy is employed (see Fig. 1) (Kricker et al. 1992; Radman 1988, 1989).

EDITING OF HOMOLOGOUS INTERACTIONS IN GENETIC RECOMBINATION BY THE COMPONENTS OF THE MISMATCH REPAIR SYSTEM

It appears that the recombination machinery contributes to the fidelity of genetic recombination mainly, if not exclusively, at the initiation stage of the strand exchange process through its requirement for MEPS-size regions of perfect homology. The major editor of genetic recombination in bacteria, and apparently in higher organisms as well, is the general mismatch repair system, particularly its mismatch-recognizing and mismatch-binding components (MutS and MutL proteins in *E. coli*) (Rayssiguier et al. 1989; Shen and Huang 1989). The multiple phenotypes of *mutS* and *mutL* mutants in *E. coli* suggest that the general mismatch repair system, which requires these *mut* genes and acts by long-patch excision and resynthesis, may be the most versatile and multifaceted DNA-editing system known. The system involves mismatch detection (MutS and, perhaps, MutL proteins) and mismatch-stimulated DNA unwinding by a helicase (helicase II, encoded by the *mutU* or *uvrD* gene). *E. coli* deficient in MutS, MutL, or MutU functions are genetically destabilized and show large increases in (1) deletions caused by replicative bypass of secondary structures (exemplified by precise and imprecise transposon excision) (Tex phenotype, Lundblad and Kleckner 1984); (2) spontaneous point and frameshift mutagenesis (mutator phenotype, hence the *mut* designation); (3) recombination of genetic markers (hyper-rec phenotype, Feinstein and Low 1986); (4) large chromosomal rearrangements resulting from crossovers between sequence-diverged repeats (chromosomal instability phenotype, Petit et al. 1991); (5) recombination with a related species such as *Salmonella* (disrupted genetic barrier phenotype, Rayssiguier et al. 1989). Each of these phenotypes results from the failure of the mismatch repair system, or its components, to act on DNA containing mismatches. The chromosomal instability, hyper-rec, and disrupted genetic barrier phenotypes all result from the failure of the defective mismatch repair system to disrupt recombinational intermediates containing mismatches in their heteroduplex regions (Rayssiguier et al. 1989; Radman and Wagner 1993).

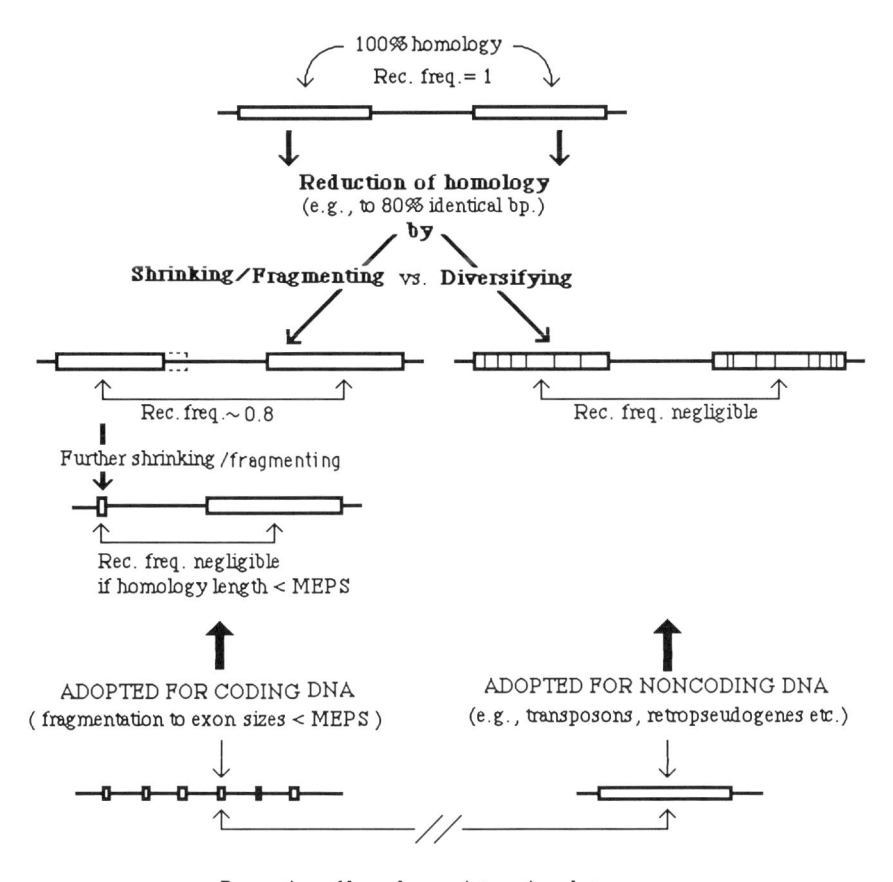

HOMOLOGOUS INTERACTIONS

Figure 1 Two strategies for avoidance of homologous recombination. The open boxes represent two blocks of identical DNA sequences. The 100% homology is reduced (arbitrarily) to 80% homology either by fragmentation or shrinking of the homologous blocks, or by 20% divergence by random point mutations. Decrease in the length of contiguous homology results in a linear decrease in recombination frequency, whereas sequence divergence results in virtual elimination of recombination. However, homology blocks shorter than the MEPS do not recombine efficiently (see text). The lower figure relates genomic elements, genes (*left*), and retropseudogenes and other repetitive elements (*right*), to the respective strategies of recombination avoidance.

The basic mechanisms (Brooks et al. 1989; Fang and Modrich 1993) and the key components of the long-patch mismatch repair system (MutS and MutL) appear to have been conserved throughout evolution (for review, see Modrich 1991; Radman and Wagner 1993). Homologs of MutS, the mismatch recognition/binding component, have been found in other bacteria, yeast, frog, mouse, and man (Reenan and Kolodner 1992a,b; New et al. 1993; I. Varlet et al., unpubl.). Not all mismatches are recognized with equal efficiency. The specificity of the long-patch mismatch repair system apparently evolved to correct errors in DNA synthesis. The most frequent polymerase errors, such as the G:T, A:C, and single base deletion/addition, are recognized and repaired most efficiently; large loops of heterology are not repaired (for review, see Modrich 1991).

Sequence divergence of repetitive sequences within eukaryotic genomes and chromosomal sequence polymorphism within populations of mammalian species are such that the only unrestricted, mismatch-free recombination is sister-chromatid exchange (SCE). SCE, but not mitotic recombination, is observed frequently in association with the repair of DNA damage. Thus, sequence polymorphism and the polymorphism-dependent action of the mismatch repair system appear to be key elements in the maintenance of chromosomal stability (Rayssiguier et al. 1989; Petit et al. 1991). For example, sequence polymorphism appears to have been the principal barrier to efficient gene targeting in mouse embryonic stem cells, a barrier that could be removed to a large extent using strictly identical, isogenic targeting DNA rather than cloned polymorphic DNA (Te Riele et al. 1992).

HOMOLOGOUS DNA INTERACTIONS OTHER THAN RECOMBINATION

Although sequence divergence is clearly a barrier to recombination, the question still remains as to how eukaryotic genomes manage to survive the creation of repeated elements, particularly the invasion of numerous copies of retrotranscribed sequences that were initially identical in sequence. A small amount of divergence may occur during reverse transcription as a result of the intrinsic imprecision of reverse transcriptases. However, a more potent mechanism for the functional and recombinational inactivation of repeated sequences is likely to have played a major role in the generation of sequence divergence of repeated sequences. Leading candidates are processes similar to the processes of MIP (Faugeron et al. 1990; Rhounim et al. 1992) and RIP (Cambareri et al. 1989), which have been observed in filamentous fungi (for review, see Selker 1990). In these organisms, almost any homologous sequence of sufficient length (greater than the MEPS) and existing in more than one copy (per haploid genome) is rapidly inactivated in the short period of haploid germ-line life after fertilization but before karyogamy, i.e., premeiotically at the dikaryon cell stage (Faugeron et al. 1990; Selker 1990). All duplicated sequences can be inactivated. In *Ascobolus immersus*, inactivation involves near to complete cytosine methylation (MIP), which is reversible (Faugeron et al. 1990), whereas inactivation in *Neurospora crassa* involves both methylation (MIP) and extensive, almost instantaneous, C to T mutagenesis (RIP) (Cambareri et al. 1989).

MIP and RIP appear to operate only within regions of sequence homology and appear always to involve at least two repeats; i.e., never is only one copy modified, suggesting that some homologous pairing process may be involved.

The results of a study of DNA-cytosine methyltransferases by Santi et al. (1983) led Selker (1990) to suggest that actual cytosine methylation may not be required for RIP mutagenesis (e.g., via deamination of 5-methyl cytosine to form thymine), but rather that mutagenesis may be an alternative to methylation. This proposal has received support from the finding that *Hpa*II methylase, in the absence of methyl-donating S-adenosylmethionine, stimulates the deamination of methylatable cytosine to uracil (Shen et al. 1992). Thus, DNA-cytosine methylases unable to methylate may become cytosine deaminases and, thereby, participate directly in RIP mutagenesis.

Evidence from studies of both plants and animals suggests that processes similar to MIP and RIP do operate in higher eukaryotes. Plants show a variety of homology-related suppressive effects on the expression of duplicated genes (for review, see Jorgensen 1990), some of which are associated with MIP-like, reversible, DNA-cytosine methylation (Matzke et al. 1993). DNA-cytosine methylation is apparently required and sufficient to block both initiation and elongation of transcription of MIP genes in *Ascobolus* (Rhounim et al. 1992). In mice, there appears to be a good correlation between the number of transgene copies and their genetic extinction caused, or at least accompanied, by cytosine methylation (Mehtali et al. 1990; M. Rassoulzadegan et al., pers. comm.). The sequences of mouse transgenes that have been silenced and methylated for more than 18 generations indicate that C to T mutagenesis in mammalian DNA repeats is much slower than RIP in *Neurospora* (M. Kricker et al., unpubl.). In addition, the spectrum of spontaneous mutations in multicopy methylated bacteriophage λ pLac transgenes in mice indicates that as many as half of phenotypic Lac mutations (analyzed in *E. coli*) are caused by CpG to TpG transitions (Gossen et al. 1991).

The results of an extensive study of the distribution of methylatable CpG dinucleotides, and TpG dinucleotides (which could result from the deamination of methylated cytosine or from cytosine deamination stimulated by the methylase), in mammalian genomes indicate that CpG deficits (and TpG surpluses) occurred in repeated DNA sequences larger than about 0.3 kb; i.e., larger than MEPS for mammalian recombination (Kricker et al. 1992). Truly unique sequences (as judged by reassociation kinetics) are CpG islands which are unmethylated and contain the expected number of CpG dinucleotides (Bird 1986). Housekeeping genes are generally single-copy genes, and their coding and noncoding sequences (introns) are relatively richer in CpGs than repeated sequences (Fig. 2). From their analysis of the CpG content versus the length of duplicated sequence, it was concluded that the minimum contiguous length of homology required for CpG loss is about 0.3 kb, which is close to the estimated MEPS for recombination in mitotic mouse cells (Bollag et al. 1989) and to recent experimental estimates for MIP in *Ascobolus* and RIP in *Neurospora* (J.-L. Rossignol and E.U. Selker, pers. comm.). Since most mammalian exons are shorter than the 0.3-kb length (Dorit et al. 1990) apparently required for MIP, RIP, and recombination, they should be relatively immune to such homologous interactions and should only interact with incompletely processed pseudogenes with which they share homology more extensive than exon (i.e., MEPS) length (Figs. 2 and 3).

CASES FROM HUMAN GENETICS

It may be that a hidden genetics of interactions between genes and pseudogenes (particularly incompletely processed pseudogenes) is the underlying cause of some human hereditary diseases. For example, mutations in the 21-hydroxylase *CYP21B* gene of many patients suffering from congenital adrenal hyperplasia have been found to be identical to those

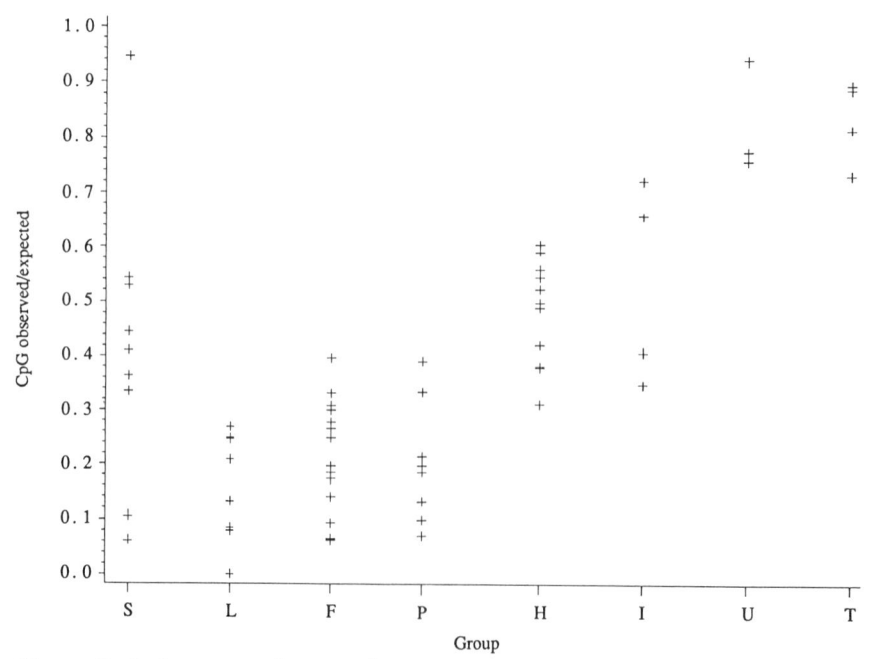

Figure 2 CpG content of repeated versus unique and unmethylated sequences. (S) SINEs, (L) LINEs, (F) members of functional gene families, (P) pseudogenes, (H) housekeeping genes, (I) unique introns, (U) unmethylated α-globin genes, (T) unmethylated transposable elements from *Drosophila* and yeast. (Data are from Table 1 in Kricker et al. 1992.)

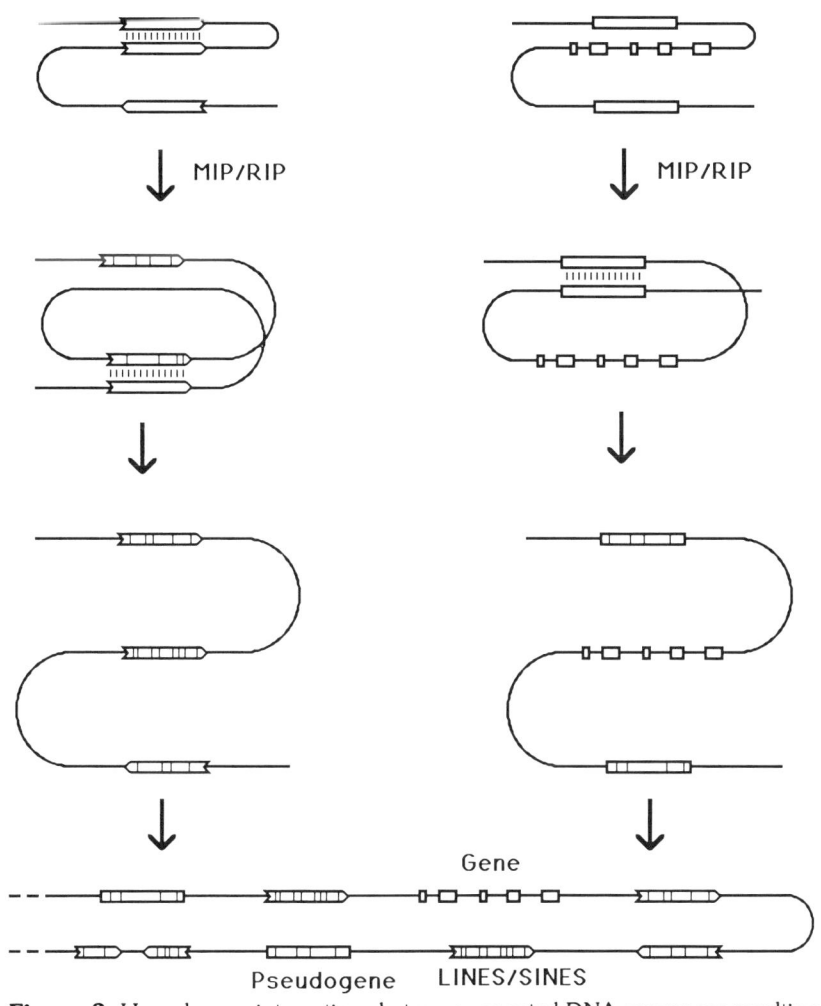

Figure 3 Homologous interactions between repeated DNA sequences resulting in DNA methylation (MIP) and mutagenesis (RIP) (see text). Transposable elements, such as retroposons, e.g., SINEs and LINEs (*left*) and processed retropseudogenes (*right*), generally share contiguous homologies longer than the minimum required (MEPS) and thus undergo accelerated sequence evolution (initially by CpG to TpG mutagenesis: vertical bars within the blocks of homology). This initial sequence divergence provides for prevention of recombination and of further MIP/RIP, thus allowing time for a slow random mutagenesis. Genes fragmented into exons shorter than MEPS are "immune" to all homologous interactions and thus preserve their coding sequence. At the bottom is a schematic presentation of the resulting chromosome structure; noncoding sequences are the principal carriers of the chromosomal sequence polymorphism.

found in the nearby pseudogene, *CYP21A*, which is unprocessed and 98% homologous to *CYP21B* (Amor et al. 1988). The presence of multiple mutations, including neutral mutations, identical to those in *CYP21A* strongly suggests that gene conversions between the gene and the unprocessed pseudogene may be the mechanism of this mutagenesis. A large fraction of human genetic diseases are associated with CpG to TpG mutagenesis in coding sequences (Cooper and Krawczak 1990) which could arise via MIP/RIP or gene conversion between genes and their pseudogenes.

The fragile X syndrome could be the result of a MIP-like phenomenon. Only in those patients with more than 200 CGG triplet repeats, i.e., a repeat block longer than the MEPS, are the sequences methylated with the coincident blockage of *FMR-1* gene transcription (for review, see Richards and Sutherland 1992).

There are cases of human gene deletions and other chromosomal rearrangements that appear to be the result of recombination between two *Alu* sequences (for review, see Hu and Worton 1992). However, the frequency of these deletions is low, particularly when considered on a per germ-line cell or per *Alu* sequence basis (there are approximately one million *Alu* sequences per diploid cell). Therefore, it may be that the average divergence of *Alu* sequences (15%), a large fraction of which are CpG to TpG changes, is the principal barrier to recombination between them.

The genetic consequences of limited sequence divergence between repeats of MEPS size can be seen in the case of human growth hormone (*GH1*) deletions which cause familial type 1A growth deficiency. The *GH1* gene is flanked by two pairs of highly homologous sequences and 48 *Alu* sequences. Among 10 independent deletions, 9 involved recombination between 594-bp repeats, which are 99% identical, and one involved recombination between 274-bp repeats, which are 98% identical. None occurred between *Alu* sequences, which are 85% identical (Vnencak-Jones and Phillips 1991).

SEQUENCE POLYMORPHISM AS A KEY ELEMENT OF GENOMIC STABILITY AND NEUTRAL EVOLUTION: A ROLE FOR "JUNK" DNA

Excessive transposon spreading and interrepeat recombination would, if unrestrained, lead to chromosomal rearrangement. Homologous interactions between genes and pseudogenes could also lead to chromosomal rearrangement and to mutagenesis as a result of crossover and gene conversion between genes by their pseudogenes. It appears that the basic strategy for the prevention of such deleterious DNA sequence interactions in mammalian genomes is itself based on homologous DNA interactions similar to MIP and RIP in fungi. The process of MIP and RIP can account for the survival of chromosomes following the invasion of "parasitic" DNA elements, because they could rapidly transform such elements into functionally and recombinationally inert junk DNA. As a consequence, the chromosomes of higher eukaryotes may be graveyards of ex-transposable elements and pseudogenes which rest in peace, thanks to RIP (Fig. 3).

It is not clear why no mechanism evolved to rid chromosomes of all useless inert DNA sequences, since they still do increase the risk of chromosomal rearrangements and impose a load on cellular machinery for their replication and maintenance. It may be that well-distributed, highly polymorphic DNA elements provide a selective advantage by serving as carriers of genomic polymorphism in the form of neutral mutations. Genomic polymorphism, as discussed above, provides a powerful barrier to genetic recombination, thus preventing expression of recessive mutations by mitotic recombination and allowing speciation by preventing fertile meioses following interspecies crosses (Rayssiguier et al. 1989; Radman 1991; Radman and Wagner 1993).

The first demonstration of the value of neutral mutations in preventing recombination came from studies with bacteria in which it was shown that both interrepeat and interspecies recombination are suppressed by sequence divergence consisting primarily of neutral mutations and that the suppression was alleviated by inactivation of components of the mismatch repair system, particularly MutS and MutL (Rayssiguier et al. 1989; Petit et al. 1991). An elegant demonstration of the prevention of homologous recombination by sequence polymorphism in higher organisms was provided by gene targeting experiments in mice. An isogenic DNA construct (*rb* gene of the 129 mouse strain) was 3 orders of magnitude more effective in homologous gene replacement than the corresponding nonisogenic (BALB/c) construct, which was diverged less than 1% by well-distributed mutations (Te Riele et al. 1992). Thus, chromosomal stability and speciation may be "phenotypes" of neutral mutations.

The need to avoid homologous DNA interactions, in order to maintain chromosomal stability, appears to have been the driving force behind the evolution of eukaryotic chromosome structure. In particular, coding regions have been fragmented into lengths too short to allow initiation of recombination, whereas noncoding regions, predominantly the junk DNA, have been sequence-diverged to prevent intra- and interchromosomal recombination, which, consequently, allows them to function as important elements in the process of speciation.

Acknowledgments

We thank Dr. John Drake for his collaboration in the sequence analysis work (performed at the National Institute of Environmental Health Sciences of the National Institutes of Health), Mikula Radman for the drawings, and the French C.N.R.S. for the support of the published and unpublished work in the laboratory of M.R. R.W. is a visiting scientist supported by C.N.R.S.

References

Amor, M., K.L. Parker, H. Globerman, M.I. New, and P.C. White. 1988. Mutation in the *CYP21B* gene (Ile-172→Asn) causes steroid 21-hydroxylase deficiency. *Proc. Natl. Acad. Sci.* **85**: 1600.

Bird, A.P. 1986. CpG-rich islands and the function of DNA methylation. *Nature* **321**: 209.

Bollag, R.J., A.S. Waldman, and R.M. Liskay. 1989. Homologous recombination in mammalian cells. *Annu. Rev. Genet.* **23**: 199.

Brooks, P., C. Dohet, G. Almouzini, M. Méchali, and M. Radman. 1989. Mismatch repair involving localised DNA synthesis in *Xenopus* egg extracts. *Proc. Natl. Acad. Sci.* **86**: 4425.

Cambareri, E.B., B.J. Jensen, E. Schabtach, and E.U. Selker. 1989. Repeat-induced GC to AT mutations in *Neurospora*. *Science* **244**: 1571.

Cooper, D.N. and M. Krawczak. 1990. The mutational spectrum of single base-pair substitutions causing human genetic disease: Patterns and predictions. *Hum. Genet.* **85**: 55.

Dorit, R.L., I. Schoenbach, and W. Gilbert. 1990. How big is the universe of exons? *Science* **250**: 1377.

Fang, W. and P. Modrich. 1993. Human strand-specific mismatch repair occurs by a bidirectional mechanism similar to that of the bacterial reaction. *J. Biol. Chem.* **268**: 11838.

Faugeron, G., L. Rhounim, and J.L. Rossignol. 1990. How does the cell count the number of ectopic copies of a gene in the premeiotic inactivation process acting in *Ascolobus immersus*? *Genetics* **124**: 585.

Feinstein, S.I. and K.B. Low. 1986. Hyper-recombining recipient strains in bacterial conjugation. *Genetics* **113**: 13.

Gossen, J.A., W.J.F. de Leeuw, A. Verwest, P.H.M. Lohman, and J. Vijg. 1991. High somatic mutation frequencies in a LacZ transgene integrated on the mouse X-chromosome. *Mutat. Res.* **250**: 423.

Haber, J.E., W.-Y. Leung, R.H. Borts, and M. Lichien. 1991. The frequency of meiotic recombination in yeast is independent of the number and position of homologous donor sequences: Implications for chromosome pairing. *Proc. Natl. Acad. Sci.* **88**: 1120.

Hawley, R.S. and T. Arbel. 1993. Yeast genetics and the fall of the classical view of meiosis. *Cell* **72**: 301.

Hu, X. and R.G. Worton. 1992. Partial gene duplication as a cause of human disease. *Hum. Mutat.* **1**: 3.

Jinks-Robertson, S. and M. Michelitch. 1993. Substrate length requirement for efficient mitotic recombination in yeast. *Mol. Cell. Biol.* **13**: 3937.

Jorgensen, R. 1990. Altered gene expression in plants due to transinteractions between homologous genes. *Trends Biotech.* **8**: 340.

Kricker, M.C., J.W. Drake, and M. Radman. 1992. Duplications-targeted DNA methylation and mutagenesis in the evolution of eucaryotic chromosomes.

Proc. Natl. Acad. Sci. **89:** 1075.
Lichten, M. and M. Fox. 1984. Evidence for inclusion of regions of nonhomology in heteroduplex products of bacteriophage lambda recombination. *Proc. Natl. Acad. Sci.* **81:** 7180.
Lundblad, V. and N. Kleckner. 1984. Mismatch repair mutations in *E. coli* enhance transposon excision. *Genetics* **109:** 3.
Matzke, M.A., F. Neuhuber, and A.J.M. Matzke. 1993. A variety of epistatic interactions can occur between partially homologous transgene loci brought together by sexual crossing. *Mol. Gen. Genet.* **236:** 379.
Mehtali, M., M. Le Meur, and R. Lathe. 1990. The methylation-free status of housekeeping transgene is lost at a high copy number. *Gene* **91:** 179.
Modrich, P. 1991. Mechanisms and biological effects of mismatch repair. *Annu. Rev. Genet.* **25:** 229.
Nassif, N. and W. Engels. 1993. DNA homology requirements for mitotic gap repair in *Drosophila*. *Proc. Natl. Acad. Sci.* **90:** 1262.
New, L., K. Liu, and G.F. Crouse. 1993. The yeast gene MSH3 defines a new class of eukaryotic MutS homologs. *Mol. Gen. Genet.* **239:** 97.
Petit, M.A., J. Dimpfl, M. Radman, and H. Echols. 1991. Control of chromosomal rearrangements in *E. coli* by the mismatch repair system. *Genetics* **129:** 327.
Puchta, H. and B. Hohn. 1991. A transient assay in plant cells reveals a positive correlation between extrachromosomal recombination rates and length of homologous overlap. *Nucleic Acids Res.* **19:** 2693.
Radman, M. 1988. Mismatch repair and genetic recombination. In *Genetic recombination* (ed. R. Kucherlapati and G.R. Smith), p. 769. American Society for Microbiology, Washington, D.C.
———. 1989. Mismatch repair and the fidelity of genetic recombination. *Genome* **31:** 68.
———. 1991. Avoidance of inter-repeat recombination by sequence divergence and a mechanism of neutral evolution. *Biochimie* **73:** 357.
Radman, M. and R. Wagner. 1993. Mismatch recognition in chromosomal interactions and speciation. *Chromosoma* **102:** 369.
Rayssiguier, C., D.S. Thaler, and M. Radman. 1989. The barrier to recombination between *Escherichia coli* and *Salmonella typhimurium* is disrupted in mismatch-repair mutants. *Nature* **342:** 396.
Reenan, R.A.G. and R.D. Kolodner. 1992a. Isolation and characterization of two *Saccharomyces cerevisiae* genes encoding homologs of the bacterial HexA and MutS mismatch repair proteins. *Genetics* **132:** 963.
———. 1992b. Characterization of insertion mutations in the *Saccharomyces cerevisiae* MSH1 and MSH2 genes: Evidence for separate mitochondrial and nuclear functions. *Genetics* **132:** 975.
Resnick, M.A., M. Skaanild, and T. Nilsson-Tilgren. 1989. Lack of DNA homology DNA in a pair of a divergent chromosomes greatly sensitizes them to loss by DNA damage. *Proc. Natl. Acad. Sci.* **86:** 2276.
Rhounim, L., J.L. Rossignol, and G. Fougeron. 1992. Epimutation of repeated genes in *Ascolobus immersus*. *EMBO J.* **11:** 4451.
Richards, R.I. and G.R. Sutherland. 1992. Dynamic mutations: A new class of mutations causing human disease. *Cell* **70:** 709.
Roeder, G.S. 1990. Chromosome synapsis and genetic recombination: Their roles in meiotic chromosomal segregation. *Trends Genet.* **6:** 385.

Santi, D.V., C.E. Garrett, and P.J. Barr. 1983. On the mechanism of inhibition of DNA-cytosine methyltransferases by cytosine analogs. *Cell* **33:** 9.

Selker, E.U. 1990. Premeiotic instability of repeated sequences in *Neurospora crassa. Annu. Rev. Genet.* **24:** 579.

Shen, P. and H.V. Huang. 1986. Homologous recombination in *Escherichia coli*: Dependence on substrate length and homology. *Genetics* **112:** 441.

———. 1989. Effect of base pair mismatches on recombination via RecBCD pathway. *Mol. Gen. Genet.* **218:** 358.

Shen, J.C., W.M. Rideout III, and P.A. Jones. 1992. High frequency mutagenesis by a DNA methyltransferase. *Cell* **71:** 1073.

Te Riele, H., E.R. Mandaag, and A. Berns. 1992. Highly efficient gene targeting in embryonic stem cells through homologous recombination with isogenic DNA constructs. *Proc. Natl. Acad. Sci.* **89:** 5128.

Vnencak-Jones, C.L. and J.A. Phillips. 1991. Hot-spots for growth hormone gene deletions in homologous regions outside *Alu* repeats. *Science* **250:** 1745.

Use of DNA Sequence Homology and Pseudogenes for the Construction of Active VSG Genes in *Trypanosoma equiperdum*

Harvey Eisen and Andrew Strand

Fred Hutchinson Cancer Research Center
Seattle, Washington 98104

African trypanosomes escape the host immune response by undergoing a process of antigenic variation whereby the variable surface glycoprotein (VSG) is periodically replaced by an immunologically distinct species. In *Trypanosoma equiperdum*, the expression of the various VSG genes is ordered. One set of complete VSG genes is expressed early after infection. The remainder (and majority) of the VSG repertoire is made by recombination of segments of pseudogenes through regions of DNA homology and site-specific recombination. It is this process that generates both the high degree of diversity and the ordered expression of the VSG genes.

Specific topics discussed include:

❑ early- and late-appearing VSGs

❑ use of DNA sequence homology to generate ordered expression of late-appearing VSG genes

❑ possible molecular mechanism for late VSG ELC formation

❑ 5'-3' joining

INTRODUCTION

African trypanosomes are extracellular hemoflagellates that parasitize animals and humans to cause African sleeping sickness. In mammals, the parasites evade the host immune system by periodically changing their VSGs, which are thought to be the unique parasite surface antigens presented to the host immune system. Although the mechanism of this antigenic variation is not yet understood, it appears that activation of the VSG genes is a complex process involving multiple steps. Silent, or basic copy (BC), VSG genes are duplicatively transposed into one of several telomere-linked expression sites where they can be transcribed as expression-linked copies (ELCs) (for review, see Donelson and Rice-Ficht 1985). Since transcription of the VSG genes can be regulated within expression sites, their insertion into an expression site, although necessary for transcription, is not sufficient (Williams et al. 1979; Pays et al. 1981; Young et al. 1983; Bernards et al. 1984; Buck et al. 1984a,b).

EARLY- AND LATE-APPEARING VSGS

The different antigenic variants do not appear randomly during the course of an infection. Rather, as has been shown for *T. equiperdum* and *T. brucei*, the variants appear at a predictable time in a loosely defined order (Van Miervenne et al. 1975; Capbern et al. 1977; Myler et al. 1985). Furthermore, transfer of any *T. equiperdum* variant to a nonimmune host results in the reappearance of the initial variant, variant antigen type (VAT)-1. Thus, the trypanosomes express their variant antigens in an order and also have a mechanism for resetting the variant clock in a nonselective environment, i.e., in the absence of antibodies. We believe that the combination of homologous recombination and selection against previously expressed VATs by antibodies can give rise to a predictably ordered sequence of VSG gene expression.

EARLY GENES HAVE COMPLETE SILENT COPIES WHEREAS LATE SILENT COPY GENES ARE DEFECTIVE

The early-appearing and late-appearing VSG genes are activated by apparently different mechanisms. VSGs appearing early in a *T. equiperdum* infection are encoded by silent copies that are complete with respect to the open reading frame (ORF) for the VSG as well as those sequences flanking the ORF that are used for integration into expression sites (Longacre and Eisen 1986). These complete silent copies can be transcribed when put into active, telomeric expression sites.

Late-appearing VSGs appear to be coded for by composite genes that exist in complete, colinear form only in expression sites. The different segments of the composite ELC genes are donated by different silent copy genes within the silent repertoire. The 3′-most 200–300 bp of the late-appearing ELCs are donated by one silent copy gene, the 3' donor. The remainder of the complete VSG gene is donated by an *unrelated* silent gene (the 5'donor) or is a mosaic composed of segments coming from two closely related members of a 5'donor gene family. The 5'donor gene families can have several members. Although there is no gross homology between the 5'and 3'donor genes, in each case thus far examined, there is a short region (75-100 bp) of homology over the region where they are joined in the ELC (Roth et al. 1986, 1989; Thon et al. 1989, 1990).

We have determined the nucleotide sequences of the silent, donor genes used for the construction of seven late-appearing *T. equiperdum* VSGs, four VSG78s and three VSG20s (Roth et al. 1986, 1989; Thon et al. 1989, 1990). Each VSG family used a single 3'donor, but these 3' donors differ between the two families. All of the 5'and 3'donor genes examined were pseudogenes, in that they do not have sufficient ORFs to encode VSGs. The silent pseudogenes contributing the 5'segments of the ELCs lack functional 3'ends. The silent genes contributing the 3' segments are mutated in their 5'regions. Thus, we believe it is likely that all composite ELCs are constructed from pseudogenes. Although the four VSG78 ELCs were all different from one another in their 5'segments, all of their nucleotide sequences could be accounted for as being derived from different segments of the same set of donor pseudogenes. Thus, all were mosaics derived from various members of the 78 5'donor family. This strongly suggests that the sequences of the ELCs were generated by recombination of the silent pseudogenes to obtain complete ORFs.

USE OF DNA SEQUENCE HOMOLOGY TO GENERATE ORDERED EXPRESSION OF THE LATE-APPEARING VSG GENES

The finding that the genes that contribute information to the late-expressed VSGs are pseudogenes, whereas, as stated above, the early-appearing VSGs are encoded by complete silent copy genes suggested the model shown in Figure 1. These silent copies have sequences flanking their ORFs (both 5'and 3') that allow them to replace any resident VSG gene in an expression site. On the 5'side, they have the conserved 76-bp repeats that precede the VSG genes in all expression sites. On the 3'side, they possess a 14-bp conserved sequence that follows the VSG ORF in all expression sites. Thus, by recombination in these conserved flanking sequences, the complete silent copies can replace any resident VSG gene in an expression site. Furthermore, such recombination leads to the production of an active VSG. Thus, complete donor genes would have a high probability of converting an expression site containing *any*

Early:

```
A B C D                C*8-14              T    ELCⁿ⁻¹
H I J K  [  X  ]       [  X  ] P Q R            BCⁿ
                       C*8-14
```

gene | conversion

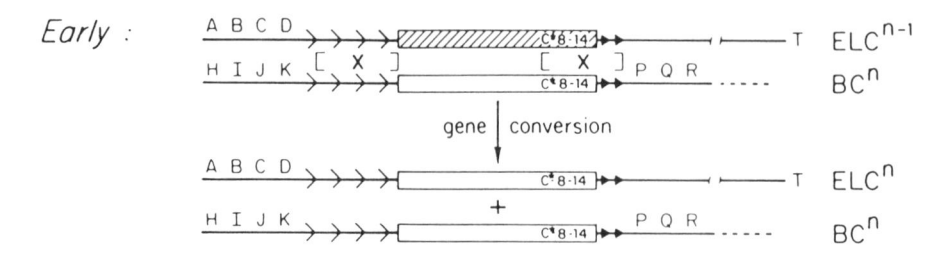

```
A B C D                C*8-14              T    ELCⁿ
        +
H I J K                C*8-14   P Q R           BCⁿ
```

Middle and Late:

```
A B C D                C*8-14              T    ELCⁿ⁻¹
H I J K    X              X     O P Q R         BC(5')ⁿ
```

gene | conversion

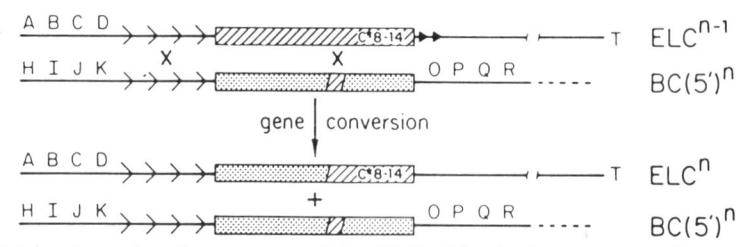

```
A B C D                C*8-14              T    ELCⁿ
        +
H I J K                         O P Q R         BC(5')ⁿ
```

Figure 1 Model for the order of expression of the VSGs. The flanking regions of early-expressed VSG BCs (BCⁿ) have some sequence homology with the expression sites, allowing transposition to occur independently of the gene occupying the expression site. Early BCs also have complete ORFs. In contrast, late-expressed BCs (BC[5']ⁿ) do not encode complete VSGs or miss the homology with the expression sites. Therefore, their expression is dependent on a recombination with the VSG gene present in the expression site at the time. Elements conserved 3' of VSG-coding regions are indicated. (C*) Region rich in cytosine; (8) TGCTACTT; (14) TGATATATTTTAAC. Large arrows represent sequences common to the 5' regions of BCs and expression sites; small arrows represent sequences common to the 3' regions of BCs and expression sites. (T) Telomere.

VSG gene, since the recombination events required do not depend on sequence homology with the resident VSG gene. These complete silent VSG genes would be used early in an infection. On the other hand, the 5' donor genes that contribute to the late VSG ELCs have defective 3' ends and are missing the conserved sequences that lie 3' to the VSG ORFs in the expression sites. We have proposed that, in order to enter an expression site, they require sequence homology with the resident VSG gene within its ORF, thus establishing a basis for ordered gene expression. The multiple events required for the formation of the composite ELCs would result in low probability of transposition and late appearance. However, where homology with the resident VSG gene existed, the particular silent gene would have a high probability of transposition (Eisen et al. 1985; Thon et al. 1990).

The model described above (Fig. 1) makes the following predictions that we have tested:

1. Early-appearing silent VSG genes can gene-convert an expression site independent of which gene occupies the site.

2. Late-appearing silent VSG genes can only convert an expression site if they have sequence homology with the current resident of that site.
3. Joining of the 5' and 3' segments of late-appearing VSG genes occurs only at the expression sites.

These predictions were tested by isolating two independent variants (VSG20* and VSG20bis) that activated the same 3' donor gene as the previously described VSG20 but used different 5' families (Thon et al. 1990). The three variants were recognized as immunologically different in mice. All of the contributing silent copies were cloned, sequenced, and shown to be pseudogenes. On the basis of homologies within the sequences of the silent copies and ELCs, it was predicted that certain variations were possible and others not possible, if the formation of complete ELCs from pseudogenes is mediated by sequence homology (Thon et al. 1990). Using PCR analysis on populations of cloned trypanosomes after passage through mice, only those variations within the 20 family that were predicted by sequence homology were found. However, variation to early VSG expression was found with all variants. Finally, variation to other late, non-20 family VSG expression was not found.

These results confirm all the predictions stated above and strongly argue that the fundamental mechanism generating the late ELCs relies on sequence homology between the resident ELC in an expression site and the "incoming" 5' (or by extrapolation, 3') donor genes. Other mechanisms, such as differences in growth rates of different variants, might well modulate this effect. However, these mechanisms can only function after the formation of the VSG genes.

A POSSIBLE MOLECULAR MECHANISM FOR LATE VSG ELC FORMATION

We propose that the formation of the late ELCs occurs in at least two steps. First, the mosaic 5' segment is produced from the relevant 5' donor pseudogenes. In a second step, this mosaic 5' segment is recombined with the resident ELC in an expression site if the proper regions of homology are present.

The extent of genetic mosaicism observed in the 5' segments of the late-appearing VSG ELCs can be very great (14 conversion patches within 750 bp in VSG78). In addition, where it has been possible to determine the number of genes that contributed to a given 5' mosaic, this appears to be two, which raises the possibility that the mosaics are made via reverse transcription. It is possible that the silent genes are transcribed at very low levels, or sporadically, and that they form dimers which are reverse transcribed with strand switching of the polymerase. This, when made double-stranded, could integrate into an expression site if it found homology with the resident gene. Processes of this type have been demonstrated for retroviruses and Ty elements in yeast (Boeke et al. 1985). More recently, it has been shown that gene conversion can occur via retrotranscription (Derr et al. 1991; Kerr and Strathern 1993). It is interesting that the African trypanosomes contain multiple copies of INGI sequences that are members of the LINES-like family of elements (Kimmel et al. 1987). Many of these have complete ORFs encoding putative reverse transcriptases. An active reverse transcriptase has been isolated from *Crithidia fasciculata* (Gabriel and Boeke 1991). Furthermore, the INGIs do not have complete gag sequences, which suggests that the reverse transcriptases might be promiscuous, since they are not confined to particles.

5'-3' JOINING

The second step in late ELC formation would be recombination of the 5' mosaic into the resident gene in an ELC. As stated above, we have not found that the 5' and the 3' donor genes contributing to a given VSG ELC belong to the same gene families. Rather, they share short regions of homology within which their joining occurs. Comparison of these sequences among donors of different VSG ELCs reveals that they are different, as expected. However, all that have been sequenced thus far contain an 11-bp sequence at, or very near, the site of recombination. This sequence, TGCAAACCCCC, can be in either orientation. In the component genes of VSG78 it is as written (Roth et al. 1986). However, in the components of VSG20 it is in the opposite orientation (Thon et al. 1989). This raises the possibility that the 3'-5' joining may actually be a site-specific recombination event. However, our results strongly argue that more extensive homology specific to the components of each gene is essential. Perhaps these homologous sequences allow for pairing of the two partners so that the putative site-specific event can occur.

WHY PSEUDOGENES?

If the role of antigenic variation in the life of the parasites is to provide escape from the host immune system, then its main function is to generate diversity. The use of defective genes guarantees the generation of diversity, since it appears unlikely that recombination of the same set of pseudogenes would give the same result twice. Furthermore, ELCs can be inactivated to become silent genes, and this would amplify the rate of evolution of the silent gene repertoire. It is likely that the complete silent genes are just such recently produced silent copies. Although the maintenance of these complete silent copy genes does not increase the diversity, it may be important in the establishment of critical numbers of the parasites in the infected mammal. Since the infecting inoculum in the wild may be small, the pseudogene mechanism for ELC generation might be too inefficient for the parasites to survive in the infected host. However, the availability of the complete silent genes would guarantee this.

Acknowledgments

We thank our colleagues Genevieve Thon and Charles Roth. This work was supported by grants from the National Science Foundation and the National Institutes of Health.

References

Bernards, A., T. De Lange, P. Michels, A. Liu, M.J. Huisman, and P. Borst. 1984. Two modes of activation of a single surface antigen gene of *Trypanosoma brucei*. *Cell* **36**: 163.

Boeke, J.D., D.J. Garfinkel, C.A. Styles, and G.R. Fink. 1985. Ty elements transpose through an RNA intermediate. *Cell* **40**: 491.

Buck, G., C. Jacquemot, T. Baltz, and H. Eisen. 1984. Re-expression of an inactivated variable surface glycoprotein gene in *Trypanosoma equiperdum*. *Gene* **32**: 329.

Buck, G., S. Longacre, A. Raibaud, U. Hibner, C. Giroud, T. Baltz, D. Baltz, and H. Eisen. 1984. Stability of expression-linked surface antigen gene in *Trypanosoma equiperdum*. *Nature*. **307**: 563.

Capbern, A., C. Giroud, T. Baltz, and P. Mattern. 1977. *Trypanosoma equiperdum*: Antigenic variations in experimental trypanosomiasis of rabbits. *Exp. Parasitol.* **42**: 6.

Derr, L.K., J.N. Strathern, and D.J. Garfinkel. 1991. RNA-mediated recombination in *S. cerevisiae*. *Cell* **67**: 355.

Donelson, J.E. and A.C. Rice-Ficht. 1985. Molecular biology of trypanosome antigenic variation. *Microbiol. Rev.* **49**: 107.

Eisen, H., S. Longacre, and G. Buck. 1985. Antigenic variation in African trypanosomes. In *The impact of gene transfer techniques in eukaryotic biology* (ed. P. Starlinger), p. 49. Springer-Verlag, Berlin.

Gabriel, A. and J.D. Boeke. 1991. Reverse transcriptase encoded by a retrotransposon from the trypanosomatid *Crithidia fasciculata*. *Proc. Natl. Acad. Sci.* **88**: 9794.

Kerr, L.K. and J.N. Strathern. 1993. A role for reverse transcripts in gene conversion. *Nature* **361**: 170.

Kimmel, B., O.K. ole-Moiyoi, and J.R. Young. 1987. Ingi, a 5.2-kb dispersed sequence element from *Trypanosoma brucei* that carries half of a smaller mobile element at either end and has homology with mammalian LINEs.

Mol. Cell. Biol. **7:** 1465.

Longacre, S. and H. Eisen. 1986. Expression of whole and hybrid genes in *Trypanosoma equiperdum* antigenic variation. *EMBO J.* **5:** 1057.

Myler, P., A. Allen, N. Agabian, and K. Stuart. 1985. Antigenic variation in clones of *Trypanosoma brucei* grown in immune-deficient mice. *Infect. Immun.* **47:** 684.

Pays, E., N. Van Meirvenne, D. Le Ray, and M. Steinert. 1981. Gene duplication and transposition linked to antigenic variation in *Trypanosoma brucei*. *Proc. Natl. Acad. Sci.* **78:** 2673.

Roth, C., F. Bringaud, R.E. Layden, T. Baltz, and H. Eisen. 1989. Active late-appearing variable surface antigen genes in *Trypanosoma equiperdum* are constructed entirely from pseudogenes. *Proc. Natl. Acad. Sci.* **86:** 9375.

Roth, C., S. Longacre, A. Raibaud, T. Baltz, and H. Eisen. 1986. The use of incomplete genes for the construction of a *Trypanosoma equiperdum* variant surface glycoprotein gene. *EMBO J.* **5:** 1065.

Thon, G., T. Baltz, and H. Eisen. 1989. Antigenic diversity by the recombination of pseudogenes. *Genes Dev.* **3:** 1247.

Thon, G., T. Baltz, C. Giroud, and H. Eisen. 1990. Trypanosome variable surface glycoproteins: Composite genes and order of expression. *Genes Dev.* **3:** 1374.

Van Meirvenne, N., P.G. Janssens, and E. Magnus. 1975. Antigenic variation in syringe passaged populations of *Trypanosoma (Trypanzoon) brucei*. 1. Rationalization of the experimental approach. *Soc. Belge Med. Trop.* **55:** 409.

Williams, R.O., J.R. Young, and P.A.O. Majiwa. 1979. Genomic rearrangements correlated with antigenic variation in *Trypanosoma brucei*. *Nature* **282:** 847.

Young, J.A., N. Miller, R. Williams, and M. Turner. 1983. Are there two classes of VSG genes in *Trypanosoma brucei*? *Nature* **306:** 196.

Index

ABL proto-oncogene, 60–61
Acholinesterasemia, 60
Acute lymphoblastic leukemia (ALL), 60–61
Acute nonlymphocytic leukemia (ANLL), 64, 70–72
Acute promyelocytic leukemia (APL), 63
ALL. *See* Acute lymphoblastic leukemia
Alu sequences, 60–61, 148
Androgen receptor, 2
Anisomorphic DNA, 112–113
ANLL. *See* Acute nonlymphocytic leukemia
Anticipation, 123, 126
APL. *See* Acute promyelocytic leukemia
Apolipoprotein gene, 62–63
Autosomal dominance, 5–6, 123

Base substitutions, 114–116
BCL2 oncogene, 63
BCR gene, 60–61
Bent DNA, 113–114

Candida, 84
CDC2 gene, 94
CDC8 gene, 94
CDC17 gene, 94
CDC21 gene, 94
chi, 63
Chromatin, 85, 92, 99
Chromosome 2, 62
Chromosome 4, 5
Chromosome 6, 5
Chromosome 9, 60–61
Chromosome-16-specific low-abundance repeats, 64–72
Chromosome 19, 5, 63
Chromosome 22, 60–61
Chronic myeloid leukemia (CML), 60–61, 63
CML. *See* Chronic myeloid leukemia
Contig mapping, 64–65
CpG islands, 123, 145–146
CpG to TpG mutagenesis, 145–148
Cruciform DNA, 108–111, 118–120
Curved DNA. *See* Bent DNA

Defined order sequence (dos) DNA, 114–121. *See also* Anisomorphic DNA; Bent DNA; Cruciform DNA; Nodule DNA; Tetraplex DNA; Triplex DNA; Z-DNA

Dinucleotide repeats, 44–47, 52–53
Dispersed simple sequence tracts
  function, 81
  genetic control of tract stability, 87–91
  and genomic instability, 80-83, 87
  structure, 81
DM. *See* Myotonic dystrophy
DM kinase, 14–15. *See also* Myotonic dystrophy
DNA. *See also* Defined order sequence DNA; Homologous DNA sequences; "Junk" DNA
  methylation, 123, 144–147
  mutation of, 114–121
  reiterative synthesis of, 129–130
  repetitive sequences, 63–72. *See also* Alu sequences; Chromosome-16-specific low-abundance repeats; Minisatellites; Telomeric repeats; Trinucleotide repeats
  secondary structure, 118
  simple sequence repeats. *See* Dispersed simple sequence tracts; Telomeric simple sequence tracts
  structure, 108–109
  supercoiling, 108, 129
  tandem repeats. *See* Dinucleotide repeats; Minisatellites
  unstable, 15–16, 27–28, 120–121
DNA helicase, 90
DNA polymerase
  I and III in yeast, 94
  reiterative synthesis mode, 129–130
  slippage
    and dispersed simple sequence tracts, 82–83, 90
    expansion, 129–130
    and simple sequence repeats, 86–87, 98–99
dosDNA. *See* Defined order sequence DNA

Escherichia coli, 86–87
est1 mutation, 93–94

Familial hypercholesterolemia, 60
FISH. *See* Fluorescence in situ hybridization
Fluorescence in situ hybridization (FISH), 66–68
FMR1 gene, 13–14, 32, 123–124, 148
Fragile sites, 62
Fragile X syndrome (FraX). *See also* *FMR1*; FRAXA; FRAXE; Trinucleotide repeats
  and dispersed simple sequence tracts, 81
  and trinucleotide repeats, 2–3, 32, 121–124, 126–127, 148
Frameshift mutations, 114–116
FraX. *See* Fragile X syndrome
FRAXA, 2–9, 13–14. *See also* FRAXE; Trinucleotide repeats
FRAXE, 3–9, 13–14. *See also* FRAXA; Trinucleotide repeats

Gene conversion, 51, 98–99, 148
Genetic counseling, 35
Genetic disorders. *See* Human genetic disorders

HD. *See* Huntington's disease
Hemophilia, 60
Homologous DNA sequences. *See also* Pseudogenes
  and genomic stability, 148–149
  and methylation induced premeiotically (MIP), 144–148
  recombination between
    editing, 142–144. *See also* Mismatch repair
    importance of, 140–141
    strategies to limit, 141–143

and repeat induced point mutation (RIP), 144–148
Human genetic disorders, 2–5, 9. *See also* Fragile X syndrome, FRAXA; FRAXE; Huntington's disease; Myotonic dystrophy; Spino-bulbar muscular atrophy; Spinocerebellar ataxia type 1
Huntington's disease (HD). *See also* Trinucleotide repeats
 age of onset, 4, 26, 30
 characteristics, 9
 and dispersed simple sequence tracts, 81
 gene product, 12–13
 genetics, 5–8
 location of defect, 5, 26–27
 phenotype, 3–4, 26
 sporadic, 31
 and trinucleotide repeats, 3, 25–36, 125–126

Imprinting, 16–17
INGI sequences, 158
IT15 gene, 125

"Junk" DNA, 59, 140, 148–149

Kennedy's disease. *See* Spino-bulbar muscular atrophy

Lesch-Nyhan syndrome, 60
Leukemia translocation breakpoints
 and Alu sequences, 60–61
 and chromosome-16-specific low-abundance repeats, 64–72
 and minisatellite repeats, 62–63
 and telomeric repeats, 61–62
 and undefined repeats, 63–72
LINEs. *See* Long interspersed nuclear elements
Linkage disequilibrium, 7–8
Long interspersed nuclear elements (LINEs), 140, 146–147, 158

Mental retardation, 3–5, 121
MEPS. *See* Minimum efficient processing segment
Minimum efficient processing segment (MEPS), 127, 141–143, 147–148
Minisatellites
 in chromosome-16-specific low-abundance repeats, 67–70
 germ-line mutations, 49–52
 hypervariability, 46–47, 63
 length-change mutations, 46–47
 and leukemia translocation breakpoints, 62–63
 localization, 45–46
 mutation rate, 46
 polarity of variation and mutation, 49
 polymorphism, 45–46
 sequence, 48
 size, 44
 unequal recombination, 43, 47, 49–51
Minisatellite variant repeat (MVR) mapping, 48–51
MIP. *See* Homologous DNA sequences, and methylation induced premeiotically
Misalignment mutagenesis. *See* Slipped mispairing
Mismatch repair, 50, 87, 90, 142–144
Multifactorial inheritance, 16
MutL, 142, 149
MutS, 142, 149
MVR mapping. *See* Minisatellite variant repeat mapping
MYC oncogene, 63
Myotonic dystrophy (DM). *See also* Trinucleotide repeats
 age of onset, 4
 characteristics, 9

and dispersed simple sequence tracts, 81
gene product, 14–15
genetics, 5–7
phenotype, 3–4
and trinucleotide repeats, 2, 33, 123–127

Neurofibromatosis, 60
Neutral evolution, 148–149
Nodule DNA, 110–111
Non-Mendelian inheritance, 5, 123. See also Anticipation

Oncogenes. See ABL proto-oncogene; BCL2 oncogene; MYC oncogene

Palindrome, 108, 121
Pericentric inversion, 64, 72
Philadelphia chromosome, 60–61, 63
PMS1 gene, 90
Pseudogenes, 140, 146–149, 155–159
Pulsed field gel electrophoresis, 64–65

Quasi-palindrome, 108, 121–122

RAD5 gene, 90
RAP1, 94–97, 99
Rapid deletion events, 96
RIF1, 96–97
RIP. See Homologous DNA sequences, and repeat-induced point mutation

Saccharomyces cerevisiae, 79–80, 87–98
SBMA. See Spino-bulbar muscular atrophy

SCA1. See Spino-cerebellar ataxia type 1
Senescence, 85–86
Sequence polymorphism, 148–149
Sherman paradox. See Anticipation
Short interspersed nuclear elements (SINEs), 140, 146–147
Simple sequence repeats. See Dispersed simple sequence tracts; Telomeric simple sequence tracts
SINEs. See Short interspersed nuclear elements
Slipped mispairing, 116–118, 129–130
Spino-bulbar muscular atrophy (SBMA). See also Trinucleotide repeats
  age of onset, 4
  characteristics, 9
  and dispersed simple sequence tracts, 81
  gene product, 12–13
  genetics, 5–8
  phenotype, 3–4
  role of androgen receptor, 2
  and trinucleotide repeats, 2, 33, 125–126
Spino-cerebellar ataxia type 1 (SCA1), 3–9, 12–13. See also Trinucleotide repeats
Strand-switch mutagenesis, 121, 130–132

TBF-α, 97
TBF-β, 97
TBF1 gene, 97
tel1 mutation, 93–94
tel2 mutation, 93–94
Telomerase, 84–85, 98
Telomere. See also Telomeric simple sequence tracts
  binding proteins, 94–99
  size control, 83–85, 92–94, 98–100
  structure, 83, 91–92
  tract addition, 84
Telomeric repeats, 61–62

Telomeric simple sequence tracts
  and cellular senescence, 85–86
  control of tract size, 83–85, 98–100
  genetic control of tract stability, 91–98
  role in chromosome stability, 80
  structure, 83, 91–92
Tetraplex DNA, 110, 112
Thymidine kinase, 94
Thymidylate synthetase, 94
Transglutaminases, 13
Translocation breakpoints. *See* Leukemia translocation breakpoints
Trinucleotide repeats
  diseases associated with, 3–5, 9, 25–36, 121–126, 148
  identification, 2–3
  instability of, 15–16, 27–28
  large-scale expansions, 10
  mutagenesis of, 126–132
  normal function, 11–12
  position, 8
  reduction, 11
  size, 8
  small-scale changes, 10
Triplet repeats. *See* Trinucleotide repeats
Triplex DNA, 108, 110–111, 120–121

Trypanosoma equiperdum, 153–155. *See also* Variable surface glycoprotein genes

Unequal recombination, 43, 47, 49–51, 82

Variable number of tandem repeats (VNTR), 43. *See also* Minisatellites
Variable surface glycoprotein (VSG) genes, 153–159
VNTR. *See* Variable number of tandem repeats
VSG genes. *See* Variable surface glycoprotein genes

X chromosome, 5, 121. *See also* Fragile X syndrome; FRAXA; FRAXE
X element, 91–92, 97

Y' elements, 91–92, 97
YAC. *See* Yeast artificial chromosome
Yeast. *See Candida*; *Saccharomyces cerevisiae*
Yeast artificial chromosome (YAC), 64–66

Z-DNA, 108–110